DNA AND GENEALOGY

Colleen Fitzpatrick, PhD
Andrew Yeiser

Rice Book Press
Fountain Valley, CA

Published by Rice Book Press

Library of Congress Control Number: 2005909527

ISBN 0-9767160-1-1

Cover designed by Kimball Clark

Printed in the United States of America

Dedicated to Michael, Quinn, and Margaret Fitzpatrick
The children of Terence Michael Fitzpatrick and Jane Snyder
The grandchildren of Emmett Martin Fitzpatrick and Marilyn Rice
The gr grandchildren of Thomas Steven Fitzpatrick, Sr. and Loretta Kelly
The gr gr gr grandchildren of John Fitzpatrick Sr. and Ellen Flynn
The gr gr gr grandchildren of Peter Fitzpatrick, Sr. and Mary Hanlon
Natives of Co. Louth & Co. Wexford, Ireland

Dedicated to Cody and Dillon Yeiser Fodness
The children of Craig Fodness and Andrea Yeiser
The grandchildren of Andrew Yeiser and Joan Leeder
The gr grandchildren of John O. Yeiser, Jr. and Gertrude Sturm
The gr gr grandchildren of John O. Yeiser, Sr. and Hettie Skeene
The gr gr gr grandchildren of George O. Yeiser and Almira Dudley Dillard
Natives of the State of Kentucky

Contents

ABOUT DNA

DNA, or deoxyribonucleic acid, is the blueprint of life, the sequence of chemicals that defines each human being as unique, with the exception of identical twins who share the same genetic makeup.

DNA is in the shape of a double helix. It gets its name from the sugar deoxyribose that forms the structure of the double helix's two spirals, which are held together by pairs of nucleotides or base pairs. There are only four nucleotides present in DNA: adenine (A), guanine (G), cytosine (C), and thymine (T). (See Figure 1.) G always pairs with C, and A always pairs with T. These pairings occur because the chemical structures of the four nucleotides prohibit any other combinations. The sequence of base pairs provides the information to produce and assemble amino acids into the proteins necessary for the life of the organism.

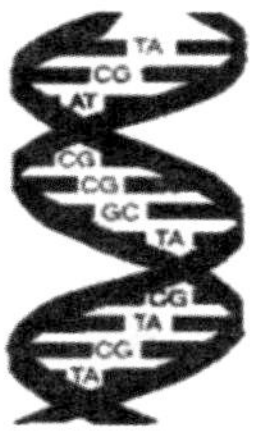

Figure 1. DNA Double helix.

There are two types of DNA, *nuclear* DNA and *mitochondrial* DNA. (See Figure 2). Nuclear DNA is present in the cell nucleus. This is the kind used by genealogists for surname studies that rely on characteristics of the DNA of the Y (male) chromosome. Mitochondrial DNA (mtDNA) is found outside the cell nucleus, and is inherited through the mother. Both Y-chromosome DNA and mtDNA are used for long term population studies and for establishing, for example, Native American or African ancestry.

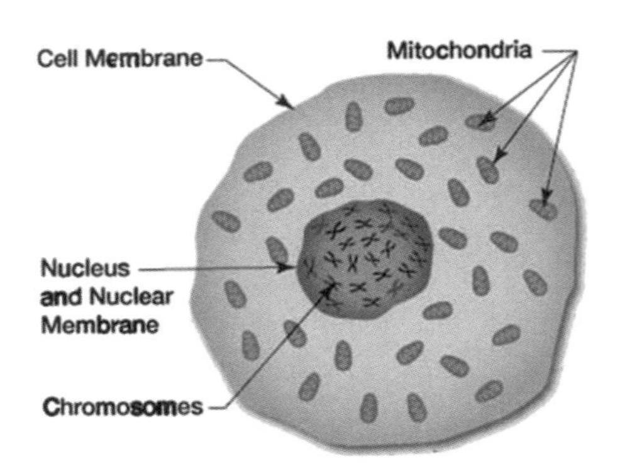

Figure 2. Nuclear (Chromosomal) DNA and mitochondrial DNA.

NUCLEAR DNA

The 3 billion base pairs of the DNA found in the nucleus of human cells are grouped in 23 chromosome pairs (see Figure 3). One chromosome of each pair is inherited from the father, and the other from the mother. Twenty-two of the pairs contain chromosomes that are functionally the same, with the same number of genes capable of the same functions. This is not the case with the 23rd pair that determines the sex of a child. A female inherits an X-chromosome from each parent, giving her a matched pair, but a male inherits an X from his mother and a Y from his father. The Y chromosome is quite different from its X partner.

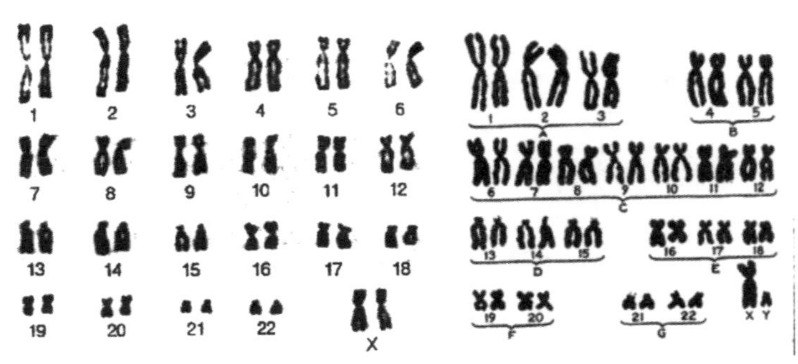

Figure 3. The 23 pairs of human chromosomes. The last pair determines the sex of a child. Females have a matched XX pair (bottom left) and males carry an XY pair (bottom right).

In a matched pair, each chromosome has genes that control the same function; for example, a gene that determines eye color. An individual can inherit a gene for brown eyes from one parent and a gene for blue eyes from the other. In this case the *dominant* gene on one chromosome

Table 1. Terminology used in genetic genealogy.

Term	Definition
Allele Value (Allele)	The number of times an STR repeats (pronounced Al-eel')
DYS Number	An identification number assigned to a DNA Y-chromosome Segment by HUGO
Haplogroup	A population group defined by specific SNP mutations
HUGO	Human Genome Nomenclature Committee
Marker or Locus (plural Loci)	Location on DNA
Mitochondrial DNA (mtDNA)	DNA found outside the cell nucleus in small organelles called mitochondria. mtDNA is inherited exclusively from the mother.
Most Recent Common Ancestor (MRCA)	The most recent ancestor shared by two people
Nonrecombinant	Sections of DNA that are always inherited from either the mother or the father. mtDNA and most of the Y-chromosome are nonrecombinant
Short Tandem Repeat (STR)	A length of DNA with a repeating sequence of chemical bases
SNP (*Snip*)	Single Nucleotide Polymorphism, a very slowly mutating location that is used to define haplogroups
Transmission Event	The transmission of a DNA marker to an offspring

PASS THE BROCCOLI

Free radicals that are one of the by-products of mtDNA energy production can damage mitochondria and surrounding cell tissue. Mitochondria do not have the repair capability that nuclear DNA has, so a mitochondrial disorder is more likely to occur spontaneously, and the distribution of defective mtDNA may vary from organ to organ. The mutation that in one person causes liver disease in another person may cause a brain disorder. The incidence of mtDNA mutations increases as the organism ages, particularly in tissues with great demand for energy.[1]

When the energy production of a mitochondrion is impaired due to damaged mtDNA, the mitochondrion may replicate at a higher rate than normal to compensate. The result is an increasing fraction of mitochondria with damaged DNA, leading to early cell death. Disruptions in energy production can be a disaster for a cell, so much so that mutations in mtDNA have been investigated as risk factors for Alzheimer's disease, Parkinson's disease, cancer, and adult-onset diabetes.[2]

Continued on p. 7

It is generally believed that 500 to 700 million years ago mitochondria were independent bacteria that colonized the precursors of the complex cells found in present-day animals and plants, forming the symbiotic relationship we observe today. A mitochondrion acts as the power plant of a cell, accepting nutrients from the cell for the production of adenosine triphosphate (ATP), the indispensable source of the cell's energy. A mitochondrion is oblong in shape. It has an outer membrane that protects the organelle and an inner membrane that is folded to increase the surface area available for energy production. mtDNA is not encoded to perform all the tasks necessary for its own proper function. A significant portion of mitochondrial function is controlled by nuclear DNA.[3]

As with nuclear DNA, mutation rates at different locations on mtDNA vary widely. Locations with very low mutation rates, called SNPs (*snips*) are useful for population studies that trace very deep roots along the exclusively female line.

When we refer to DNA or to the human genome, unless stated otherwise, we are not referring to mtDNA. We are referring to chromosomal DNA found in the nucleus. Terminology used in genetic genealogy is given in Table 1.

Broccoli (cont)

Enzymes produced by the cell scavenge free radical toxins and slow the damage they cause. Research has shown that severely restricting food intake can prolong the life of laboratory mice by as much as 50%, presumably because the level of free radicals produced is kept below the removal capacity of the scavengers. Eating foods such as broccoli that contain antioxidants can slow the aging process by supplementing the cell's natural supply of scavengers to remove free radial waste products. So make sure you eat your broccoli. Your life depends on it!

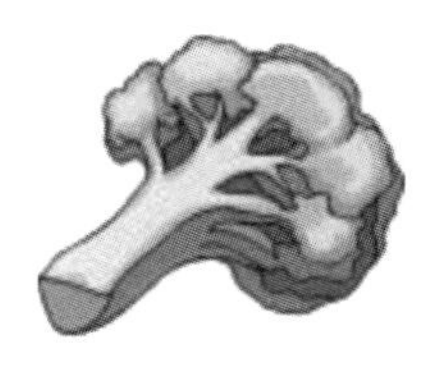

1. Wei, Y. H. et al, "Mitochondrial theory of aging matures - roles of mtDNA mutation and oxidative stress in human aging," Zhonghua, Yi Xue Za Zhi (Taipei),**64**(5) 259-70, May 2001.
2. http://www.genomenewsnetwork.org/articles/04_02/mito_dna.html

THE HUMAN GENOME PROJECT

The sequence of base pairs comprising DNA is analogous to a book printed with one long string of seemingly random letters of the alphabet. A gene located somewhere along the chromosome is like a word appearing somewhere in a string of random letters. The challenge of locating genes to decipher the human genome is equivalent to distinguishing random gibberish from meaningful content. This is a difficult process, especially when very little is known about the language the words are written in, including typical word lengths and the combinations of letters that can be used as their syllables.

The Human Genome Project (HGP) was officially launched in 1990 as a joint effort between the U. S. Department of Energy and the National Institutes of Health. The project was planned as a $3B, 15-year effort (a cost of about $1 per base pair). The goals of the HGP were primarily to determine the sequence of its 3 billion chemical base pairs and to identify all the 20,000 to 25,000 genes in human DNA. The data accumulated were to be stored in databases for public access, and the technologies developed were to be licensed to the private sector to stimulate the growth of the

Table 2. The Human Genome[1]

Human DNA contains 3,147 million chemical nucleotide bases (A, G, C, T). The total length of DNA present in one adult human is about 2×10^{13} meters, the distance from the earth to the sun and back. The DNA in an average chromosome is about 5 cm long.
The average gene consistes of 3,000 bases, but sizes vary greatly, with the largest known human gene consisting of 2.4 million bases.
The total number of genes is estimated at 30,000 - much lower than previous estimates of 80,000 and 140,000 that had been based on extrapolations from gene-rich areas as opposed to a composite of gene-rich and gene-poor areas.
Almost all (99.9%) nucleotide bases are exactly the same in all people; and 97% are the same as the chimpanzee's.
The functions are unknown for over 50% of identified genes.
Repeated sequences that do not code for proteins (junk DNA) make up at least 50% of the entire human genome.
Genes appear to be concentrated in random areas along the genome, with vast expanses of nonfunctional junk DNA in between.
Chromosome 1 has the most known genes (2,968), and the Y-chromosome has the fewest (231).
Unlike humans' seemingly random distribution of gene-rich areas, many other organisms' genomes are more uniform, with genes evenly spaced throughout.
Scientists have identified about 1.4 million locations where single-base DNA differences occur in humans.

1. Human Genome Project Information, "About the Human Genome Project", http://www.ornl.gov/sci/techresources/Human_Genome/project/about.shtml

biotechnology industry.[4] The social, legal, and ethical implications of the project were to be addressed.

In 1998, a second effort was initiated by a venture capital-funded company called Celera Genomics (founded by Craig Ventner) to sequence the human genome using a faster method that cost about one tenth of that used by the government-funded HGP.[3] Because of the extent of international cooperation on the project, the rapid advance of DNA sequencing technology, and the growth of computational capacity, the HGP was completed two years ahead of schedule. HGP scientists and Celera Genomics jointly announced in April 2003 that 99% of the human genome had been sequenced to 99.9% accuracy. All humans have unique gene sequences, therefore the data published by the HGP does not represent the exact sequence of every individual's genome.[5] It is the combined genome of a small number of anonymous donors.

Only about 50% of the genes in the human genome have been identified so far.[6] There is a vast amount of research being done to identify the rest. There is also much work being done in the area of functional genomics, understanding the role that genes play in sustaining and defining life. This includes the effects of genetic mutations that cause diseases. Facts about the human genome are listed in Table 2.

The sequencing of the human genome has provided a foundation for researching biological systems over the next many decades. The genome of the bacillus *Haemophilus influenzae* was the first genome of a freestanding organism to be sequenced. Its genome consists of 1,830,140 base pairs of DNA and contains 1,740 genes.[7] The sequencing was completed in 1995 at the Institute of Genomic Research, a nonprofit research organization in Bethesda, MD.[8] There are many groups now working on sequencing the genomes of other living organisms, including many mammals, reptiles, amphibians, and fish. The genomes of many plants and types of bacteria have also been sequenced. For a comprehensive list of sequencing projects see the National Center of Biotechnology Information's website at http://www.ncbi.nlm.nih.gov/. As of August 2005, the public collection of genomic sequences worldwide contained the individual genes and partial and complete genomes of over 165,000 organisms for a total of over 100 billion bases. This is slightly less than the number of stars in the Milky Way galaxy.[9]

THE BIRTH OF DNA FINGERPRINTING

DNA analysis is used today in countless applications–from paleo-DNA analysis of extinct animal species, to exonerating inmates on death row, to the identification of Anna Anderson as an imposter of the Tsar's still-missing daughter Anastasia, to the use of DNA for anti-counterfeiting applications to gene therapy for curing diseases. DNA is in the news almost every day. But where did this revolution start? And how did DNA analysis become such a valuable tool for genealogists?

DNA fingerprinting was developed in 1984 by Sir Alec Jeffreys and his group at Leicester University in England. They were searching for regions in human DNA with genetic signatures that differed enough from person to person to be useful to map genes and to develop diagnoses for inherited diseases.[10] It was already known that variations in DNA called SNPs (Single Nucleotide Polymorphisms, or *snips*) were inherited. But SNPs mutate extremely slowly, so that very large portions of the population have the same SNP. SNPs are not very useful in applications depending on a large variation in a human genome.

DNA IN THE NEWS

DNA is in the news almost every day. Some applications of DNA have become routine, such as the type of DNA analysis that is used in forensic science for investigating crimes. Paternity testing has also become commonplace.

As more is learned about DNA, new applications are emerging that are not familiar to the public. Some are highly controversial, while others use well-known techniques applied in innovative ways. Getting familiar with some of the frontiers of DNA analysis and its applications is fun.

The focus of Sir Alec's efforts was on regions of DNA called minisatellites where a sequence of nucleotides between 10 bp (base pairs) and 100 bp long stutters, that is, it is repeated as many as several hundred times.[11] Since the first minisatellite, or marker, was discovered in 1980 by Wyman and White, many other such markers had been found. Sir Alec was looking for a way to identify them more reliably and to isolate many minisatellites at one time. In successfully doing so with the myoglobin gene that produces the red color in muscles, he noticed that the collection of minisatellites he had isolated varied widely from one individual to the next. At 9 a.m. GMT, September 15, 1984, the science of DNA fingerprinting was born.

The first application was in 1985, when DNA fingerprinting was used to resolve a two year old immigration dispute. A teenage boy who was a naturalized citizen of the UK had been denied re-entry into England after leaving the country to visit his family in Ghana. On return his passport appeared to have been tampered with, so that authorities could not be sure whether another boy had been substituted in his place. Conventional blood typing showed that there was a 99% chance that he was related to the woman who claimed to be his mother, but it could not rule out that he was her nephew, the son of her sister in Ghana.

The situation was complicated by the fact that the woman was uncertain of the identity of the father of the boy, although she knew the father of her other three children. To determine the parentage of the boy, Sir Alec first reconstructed the DNA profile of the father of the other three children by finding which of the features in

GENE THERAPY

Genes get a lot of PR, but it's the proteins they specify that are responsible for essential life-sustaining processes. Proteins constitute the majority of cellular structure and control cellular processes. When a gene mutates, the protein that should be modeled by that gene either is not produced or it is a modified version of the one the organism must have to function properly. The result may be a genetic disorder such as cystic fibrosis or diabetes.

Gene therapy involves replacing a defective gene with a healthy version, a kind of "gene surgery." The most common form of gene therapy relies on a vector, or carrier, to transport the good copy of a gene to a patient's DNA. The most common vectors used are viruses since they have the ability to encapsulate genetic material and deliver it through cell walls. The viruses are modified to prevent an immune reaction.

Continued on p. 12

Gene Therapy (cont)

On September 14, 1990, the first gene therapy trial was conducted at the University of Southern California Medical Center. Ashanti DeSilva, a 4-year old girl from the Cleveland area, was treated with gene therapy for an inherited disorder known as adenosine deaminase (ADA) deficiency that severely compromised her immune system. ADA had recently been approved for gene therapy by the FDA. It is caused by a single genetic mutation. Ashanti was injected with her own white blood cells that had been genetically modified to carry a healthy version of her defective gene. For the first time, a disease was treated by modifying the genetic structure of the patient.[1]

Ashanti's treatment was a success, sparking anticipation of a medical revolution. But since then, gene therapy has had mixed results. In 1999, Jesse Gelsinger became the first documented patient to die from gene therapy treatment, rather than from the disease he was being treated for. Many other clinical trials have failed.

Continued on p. 13

their collective profile were absent in their mother's profile. He assumed that what remained were characteristics of their common father. These were subtracted from the boy's profile. The remaining features were all present in the mother's DNA. Sir Alec was not only able to prove that the boy was the woman's son with a 99.997% probability, he was also able to show that all four children had the same father.[12]

Many applications of DNA to human identification quickly followed this success. In 1986, the first use of DNA fingerprinting in a forensic investigation involved the rape and murder of two teenage girls in Enderby, England. Authorities apprehended a kitchen porter at a local mental institution who confessed to the second murder, but refused to take responsibility for the first. Attempting to prove that the suspect had committed both murders, police turned to Dr. Jeffreys and his new technique. The results of comparing the DNA extracted from semen found at the two crime scenes with DNA from the suspect clearly showed that the crimes were committed by the same man–but not the man they had in custody. On November 21, 1986, legal and forensic history was made when the kitchen porter became the first accused murderer to be found innocent on the basis of DNA fingerprinting.[13]

In the first DNA manhunt, the Enderby police tested every local man between the ages of 16 and 64 years old. Unfortunately, testing over 4,500 men produced no match to the murderer's DNA. Shortly afterwards, a woman reported to the police a conversation she heard in a local bar in which a man mentioned that he had been bullied into taking the DNA test in the place of a coworker using a fake passport. The coworker was detained and became the 4,583rd and last man to be tested. He was a match. History was made again when this man, Colin Pitchfork, became the first person to be convicted of a crime based on DNA evidence.

Gene Therapy (cont)

Yet there are still patients who benefit from gene therapy. As of 2000, 10 children with X-linked severe combined immunodeficiency disease (X-SCID) apparently have been cured by the treatment. SCID is a disorder that destroys the immune system, so that SCID children must live their lives in sterile isolation. SCID is the disease from which David, the Boy in the Bubble, suffered. Because he was incapable of fighting infection, David lived his entire life in a plastic enclosure, never even having physical contact with his mother.

1. Dawn Vargo, CitizenLink, Focus on Social Issues, Bioethics/June 23, 2004, http://www.family.org/cforum/fosi/bioethics/genetics/a0032608.cfm

For DNA fingerprinting to be put into common use, the technique had to be simplified. The testing of many thousands of samples during the Enderby investigation had been a time-consuming and laborious process, with a single test often requiring several weeks. The highly complex DNA profiles produced by the technique were also impossible to decipher for DNA that contained more than one person's genetic material. To overcome these limitations, Sir Alec composed a test panel of only a few highly variable minisatellites that could be detected individually, rather than as part of a large group. This reduced the complexity and increased the speed of the analysis, permitting automation for efficient testing of a large number of samples—and very importantly made the results compatible with databasing. If only one such minisatellite were tested, the results would match so many people that the technique would be useless. But by using a panel of minisatellites, the chances that the resulting DNA profile would match only one person approach 100%.

A second innovation in 1983–1988 that made it possible to use DNA fingerprinting on a large scale was the development of the Polymerase Chain Reaction (PCR) technique of DNA amplification. The PCR technique made it convenient to test very small amounts of DNA typically harvested from cells collected from the inside of the cheek with a swab, or found in human remains and on cigarette butts and coffee cups. However, the PCR technique is limited in the number of base pairs it can amplify (at that time, 1000 to 2000 bp) and many minisatellites are much longer.[14] For PCR to be an effective technique for DNA analysis, it was necessary to use newly identified microsatellites, also called Short Tandem Repeats (STRs), that were only 1 to 7 bp long, had only 5 to 100 repeats, and yet still showed a person-to-person variation similar to that of minisatellites. With all these elements in place, the stage was set for DNA analysis to become a powerful tool for genealogists.

THE GENETIC GENEALOGY REVOLUTION

During conception, the sperm and the egg contribute equally to the formation of a zygote (fertilized egg), with two exceptions. The sperm does not contribute its mtDNA, and half of the time the sperm contributes a Y-chromosome, creating a male.[15] The other half of the time it contributes an X, creating a female. The egg always contributes an X, so that males have an XY chromosome pair and females have an XX pair.

All children receive mtDNA from the mother, but only males inherit a Y-chromosome, and that can only come from the father. These two types of DNA are nonrecombinant, that is, they remain virtually unchanged as they are passed down the generations with the exception of very occasional mutations. These mutations are used to track exclusively female and exclusively male genetic heritage.

Mitochondria

Mitochondria carry their own DNA, different and separate from the DNA found in chromosomes in the cell nucleus. mtDNA is a circular structure only 16,569 bp long. It was first sequenced in 1981, creating the Cambridge Reference Sequence (CRS).[16] In a slightly modified form, the revised CRS[17] is used as a master template against which to compare all other mtDNA profiles.

Mitochondrial DNA is divided into the coding and the control regions that have different mutation rates. A mutation occurs somewhere in the coding region on the average every 5,138 years; a mutation occurs in the control region on the average every 20,180 years. The slow mutation rate of the control region (also called the D-loop) is useful for defining deep female ancestry. The higher mutation rate of the coding region allows more detailed analysis of maternal lineages. The control region was used by Bryan Sykes in his book *The Seven Daughters of Eve*, where he showed that 98% of all people of Western Europe descend from only seven women who lived in Europe within the last 50,000 years.[18] These "clan mothers" themselves descend from a common ancestor "Mitochondrial Eve" who lived in Africa about 100,000 to 200,000 years ago.[19] In 2000, in response to the great success of his book, Dr. Sykes started Oxford Ancestry to offer mtDNA testing to the public.

HELLO DOLLY!

There has been much debate in the media about human cloning. But chances are, we all have met human clones. Human clones have been among us for a long time. We call them identical twins.

There are three kinds of cloning: recombinant cloning, reproductive cloning, and therapeutic cloning.[1] Recombinant cloning involves the transfer of a DNA fragment from a source organism to self-replicating genetic material such as a virus. When the virus is introduced into a recipient cell, it multiplies along with the cell. This is the principle of gene therapy.

Reproductive cloning involves the creation of a genetic twin of an organism by transferring the nucleus of one of its cells into the egg cell of another organism from which the nucleus has been removed. The reconstructed egg containing the DNA from a donor cell must be treated with chemicals or electric current to stimulate cell division. After it reaches a suitable size, the donor egg is implanted into a female where it can continue to develop.

Continued on p. 16

Dolly (cont)

This method was first used successfully in 1997 at the Roslin Institute in Scotland to create a sheep named Dolly, the first mammal to be cloned from adult DNA. Because the clone retains mtDNA of the host cell, it is not identical to the donor. But reproductive cloning is the closest we have come in the laboratory to creating identical twins.

The third kind of cloning is therapeutic cloning. It is at the heart of what the media call the "Stem Cell Debate." The goal of therapeutic cloning is not to create a fetus that can be brought to full term, but rather to create one or more embryos from which stem cells can be harvested at the very early stages of development. Such harvesting destroys the embryo, prompting debate on the ethics of the technique. Stem cells retain the ability to differentiate into other cell types. This allows them to act as a repair sys-

Continued on p. 17

Short Tandem Repeats (STRs)

Y-chromosome Short Tandem Repeat (STR) microsatellites mutate much more rapidly than mtDNA, and on a time scale that makes them useful for genealogy. A typical Y-STR genealogy test panel consists of relatively slow and fast mutating markers, with an average mutation rate of about 4/10% (four tenths of one percent or 0.4%) per marker per generation, so that a mutation is expected on a marker on the average every 250 years.

The first suggestion of using Y-STRs for genealogy was made by Mark A. Jobling and Chris Tyler-Smith in their 1995 paper "Fathers and sons: the Y-chromosome and human evolution."[20] At that time, Y-DNA marker identification lagged behind mtDNA research primarily because of the high complexity of the Y-chromosome. Y-markers were hard to find, and few populations had been adequately surveyed.

Yet the length and complexity of the Y-chromosome offered a rich library of marker types and mutation rates that would allow Y-DNA to be used to study ancestry on a variety of time scales. Not only did Y-chromosome markers offer potential for deep-ancestry research along the exclusively paternal line on the same time scale as that offered by mtDNA markers along the exclusively female line; they also promised a new, highly innovative method that could be used by genealogists for tracing more recent ancestry. Since a family name is typically handed down along the male line in western societies, studying Y-chromosome STRs is useful to a genealogist for tracing the history of the exclusively male lines of his family.

Dolly (cont)

tem for the body, replenishing other cells as long as the organism is alive.[2] Stem cells may have the potential for organ repair and for treating diseases.

1. Human Genome Project Information, http://www.ornl.gov/sci/techresources/Human_Genome/elsi/cloning.shtml
2. http://en.wikipedia.org/wiki/Stem_cells

Thomas Jefferson

One of the earliest and most popularized DNA studies of a surname was that of Thomas Jefferson. There had always been rumors supported by some historical evidence that Jefferson had fathered several children with his quadroon house slave, Sally Hemmings. It was known that all of Sally's proven children were conceived while Jefferson was at his home in Monticello in Virginia, and that when Jefferson died, these children were released and Sally was allowed to live as a free woman of color.[21] Analysis performed by *E. A. Foster et al*[22] found an exact match between Y-chromosome DNA of a descendent of Jefferson's uncle Field Jefferson and that of a descendent of Eston Hemmings, Sally's youngest son. While this only showed that Eston was descended from a Jefferson through an exclusively male link, considering the circumstantial evidence, it seems certain that the connection was to Thomas Jefferson himself.

The analysis disproved two other claims, that Sally's children were offspring of Jefferson's brother-in-law Peter Carr (husband of Jefferson's sister), and the claim made by the descendents of Thomas Woodson that he was the oldest of Sally's sons by Jefferson. Both the Carr and the Woodson DNA profiles were very different from that of Jefferson, ruling out these possible connections.[23]

The Revolution

The first steps to establishing a genetic genealogy community were achieved in 1999 when Bennett Greenspan approached Mike Hammer at the Laboratory of Molecular Systematics and Evolution at the University of Arizona about doing testing on the Nitz family name. Greenspan had recently met a roadblock in following the paper trail of his mother's mother's father's family. He had located a group in Buenos Aires, Argentina with the same unusual surname who apparently emigrated from the same area of the Crimea three years after his mother's family arrived in Omaha, NE. Examination of Russian exit passports produced nothing beyond circumstantial evidence that the two families might be related.

Greenspan remembered a story he had seen in 1997 of a DNA study performed at Hammer's laboratory that showed a common ancestor for the Cohanim, the Jewish priesthood.[23] He also recalled the 1998 Jefferson/Hemmings study. He reasoned that if DNA testing could be used to solve the impasse in his own personal research, other genealogists would be interested in using it for similar purposes. Within weeks, Greenspan and Hammer launched a proof of concept study on 24 individuals to test the viability of their idea. The results were exciting.

With Greenspan handling the business end, and the Hammer Laboratory providing DNA testing services, Family Tree DNA (Family Tree DNA), became the first company to provide DNA testing for genealogy. When Family Tree DNA started accepting orders in April 2000, the genetic genealogy revolution was on!

REFERENCES–ABOUT DNA

1. http://www.jdaross.mcmail.com/mitochon.htm
2. See http://www.newscientist.com/news/news.jsp?id=ns99992716 or *The New England Journal of Medicine*, 347, p. 576 for a rare exception.
3. http://en.wikipedia.org/wiki/Mitochondrial_disease
4. http://en.wikipedia.org/wiki/Human_genome_project.
5. Ref. 3, *op cit.*
6. http://www.ornl.gov/sci/techresources/Human_Genome/project/info.shtml#draft
7. http://en.wikipedia.org/wiki/Haemophilus_influenzae
8. The Institute for Genomic Research, http://www.tigr.org/

9. The U. S. National Library of Medicine Press Release, "Public Collections of DNA and RNA Sequence Reach 100 Gigabases", August 22, 2005. http://www.nlm.nih.gov/news/press_releases/dna_rna_100_gig.html.
10. https://sciencegrants.dest.gov.au/SciencePrize/Pages/Doc.aspx?name=previous_winners/Aust1998Jeffreys.htm.
11. http://en.wikipedia.org/wiki/Minisatellite
12. http://www.forensicmag.com/articles/0904dna.asp?pid=17&articleText=0904dna and http://arbl.cvmbs.colostate.edu/hbooks/genetics/medgen/dnatesting/immigrant_dna.html and http://www.w3ar.com/a.php?k=980.
13. *The Casebook of Forensic Detection: How Science Solved 100 of the World's Most Baffling Crimes*, Colin Evans, John Wiley and Sons, Inc., 1996, p. 61.
14. The Wellcome Trust, "DNA Fingerprinting and National DNA Databases", 24/02/2004, http://www.wellcome.ac.uk/en/genome/genesandbody/hg07f007.html.
15. M. A. Jobling, C. Tyler-Smith, "Father and sons: the Y-chromosome and human evolution", *TIG*, **11** (11), November 1995, pp. 449 – 456.
16. S. Anderson, A. T. Bankier, B. G. Barrell, et al. (14 co-authors), "Sequence and organization of the human genome", *Nature* **290**:457-465, 1981.
17. R. M. Andrews, I. Kubacka, P. F. Chinnery, R. N. Lightowlers, D. N. Turnbull, N. Howell, "Reanalysis and revision of the Cambridge reference sequence for human mitochondrial DNA", *Nat. Genet.* **23**:147, 1999.
18. http://www.amazon.com/exec/obidos/tg/detail/-/0393323145/102-7342205-7108912?v=glance.
19. R. L. Cann, M. Stoneking,A. C. Wilson, "Mitochondrial DNA and human evolution", *Nature* **325**(6099);31-6, Jan 1987 and M. Stoneking, "Mitochondrial DNA and human evolution", *J. Bioenerg. Biomembr* **26** (3);251-9, Jun 1994.
20. Ref. 12, op cit.
21 Annette Gordon-Reed, *Thomas Jefferson and Sally Hemmings: An American Controversy* (University of Virginia Press, Charlottesville, 1998) as discussed by Thomas H. Roderick, "The Y-chromosome in Genealogical Research: 'From Their Ys a Father Knows His Own Son'", *National Genealogical Society Quarterly* **88**, p. 122-143, June 2000.
22. E. A. Forster et al, "Jefferson Fathered Slave's Last Child," *Nature* **396** 1998:27-28.
23. Thomas H. Roderick, "The Y-chromosome in Genealogical Research: 'From Their Ys a Father Knows His Own Son'", *National Genealogical Society Quarterly* **88**, p. 122-143, June 2000.
24. M. J. Hammer et al, "Y-chromosomes of Jewish Priests" *Nature*, **385**, 1997.

THE INS AND OUTS OF GENETIC GENEALOGY

Just because you are investing in a DNA test doesn't mean you should throw out all those notes you've been accumulating on your family tree. When used together, DNA and written documentation offer complementary tools for researching family connections. If they exist, documents can record exact relationships. Yet documentation can be misleading and contain erroneous information. In contrast, DNA can only tell you that two people are related. It cannot tell you the exact nature of the relationship, but DNA cannot be destroyed or changed, and it is never "wrong."

WEIRD DNA

We can sometimes learn more by studying exceptions than by studying rules. There are countless examples in science where a single inconsistency revealed far more than what had been learned from a large body of consistent data.

There are rare situations involving "weird DNA" that fall into this category. Read more about chimeras, mosaics, sickle cells, and more in the sidebars!

PALEO-DNA

Paleo-DNA is obtained from the remains of ancient people and animals. Analysis can be a challenge, as DNA deteriorates upon the death of an organism, and DNA from microbes that attack the remains contaminate what is left. A sample can also be contaminated by fossil-hunters and archeologists. Normally only a small fraction of the DNA recovered from remains is that of the species of interest. The majority of the sample is a "soup" of DNA from microbial contaminants.

It is the challenge of paleo-DNA analysts to identify and isolate DNA fragments in the soup that belong to the species. This can be difficult with nuclear DNA. In principle, it is possible to do so by comparing all the fragments in the soup to the species' genome. But in practice this is very difficult, because of the small size of the scraps that survive compared to the large size of the nuclear genome.

Using mitochondrial DNA offers a much higher probability of success. The size of the mitochondrial genome is only a tiny fraction of that of the nuclear ge-

Continued on p. 23

In the absence of a paper trail, DNA is your most valuable guide directing your research. If a person's DNA profile (also called his haplotype) closely matches that of someone already known to belong to a family line, the person can focus his efforts on finding a link to that person's family line. If a person's DNA profile matches someone's who does not have a confirmed pedigree, they can pool their efforts to find out how they are related to each other. Each may have part of the story.

If the person's haplotype is very different from everyone else's in his surname group, he could link to the family is through a so-called "non-paternity event," that is, an adoption, a name change, or an illegitimacy. By investigating family folklore, he may find the surname he is genetically linked to, or he can look for a match with someone with a different surname whose DNA results are in an online database.

The Mumma surname study provides an interesting example of such a situation.[1]

"A man with the surname Bell contacted the Mumma Project Administrator, relating the story that his great grandfather who was a traveling salesman in Ohio was "robbed and mur-

dered" before he could marry his great grandmother. She was pregnant with his grandfather at that time. His great grandmother eventually married a man named Bell, who adopted the "illegitimate" boy. The family oral tradition claimed his great grandfather's name was "Elmer Maumau" but there was uncertainty because no written records exist. An Elmer Mumma was in the Mumma family database, but he had not been murdered, had married and had raised a family. It didn't appear as if he was a candidate.

"Being curious all these years as to his true 'genetic' surname, he submitted a sample of his DNA to Family Tree DNA_for analysis.

"His haplotype matched exactly three descendants of immigrant Peter Mumma. It was concluded that his true genetic surname is Mumma. These results were presented to one of the great grandchildren of Elmer Mumma who revealed that Elmer was a traveling salesman, he was single at the time Mr. Bell's grandfather was conceived, and he had been living in same area of Ohio. After comparing all of the facts and even family photographs, it was concluded that Elmer Mumma was the father of Mr. Bell's illegitimate grandfather."

DNA testing can help correct mistakes in written genealogies. Another example from the Mumma study involves

Paleo-DNA (cont)

nome, and the large number of mitochondria present in each cell provide more genetic material so that a higher fraction of the genome may be represented by the fragments. However, because mtDNA mutates much more slowly than nuclear DNA, using mtDNA can only provide broad information about the genetic relationship between an extinct species and its modern day counterparts.

Recently scientists at the U. S. Department of Energy's Joint Genome Institute were able to sequence nuclear DNA from 40,000 year old bones of two cave bears *(Ursus spelaeus)* found in the Austrian Alps.[1] This was not accomplished by new DNA sequencing techniques, but by relying on increases in computational power to identify the small fraction of the sample that was cave bear DNA.

After sequencing all the DNA recovered, scientists compared the fragments to the dog genome, which is publicly available. Dogs and bears diverged 50 million years ago, but still share 92% of their genome.[2] The computer was

Continued on p. 24

Paleo-DNA (cont)

able to find the needle in the haystack–six percent of the sample was undamaged cave bear DNA. Although enough DNA was recovered to sequence the whole cave bear genome, only fragments of 21 genes were examined as a proof of concept. The data showed that the cave bear, extinct for 10,000 years, is more closely related to the modern-day brown bear than to the black bear.

The cave bear was chosen for the experiment because cave paintings show that the

bears lived at the same time as Neanderthals. Successful analysis of cave bear DNA suggests that DNA of Neanderthals dating to about the same era may be sequenced.[3] Cave bear DNA also ruled out the possibility of contamination, since not many cave bears work in DNA sequencing laboratories.

The best seller, *Jurassic Park,* was based on the idea of cloning dinosaurs from DNA found in dinosaur blood from

Continued on p. 25

an ambiguity caused by the existence of two men both named Christian Mumma, both born around the same time, and who both lived in the township of Lancaster, PA. One Christian was the grandson of Jacob Mumma who arrived in the U.S. in 1731; the other was the grandson of Leonard Mumma who immigrated to the U. S. in 1732. Although two Mummas tested for the study had been told they were descendents of the Christian who belonged to Jacob's line, they clearly showed the mutations of Leonard's branch of the family. When the written pedigrees of these Mummas were investigated more closely, an uncertainty in the records was discovered that had caused a genealogist to assign the Christians to the wrong family lines. Thanks to DNA testing, the family histories were corrected.

SO HOW DOES DNA WORK?

When a cell divides, its DNA replicates to provide each daughter cell with a copy. DNA does this by unzipping its double helix, resulting in two single strands of DNA with complementary nucleotide sequences. Each strand then attracts the chemicals necessary to rebuild its missing half (Figure 1).

Figure 1. DNA replication.

In most cases, the reconstructed DNA is identical to the original, but occasionally a mistake or mutation occurs during the replication process, so that the copied DNA is slightly different from the original. Mutations are natural phenomena that occur in all types of DNA. Mutations occur with a statistical frequency, that is, only the probability of a mutation can be predicted. Most mutations have no effect, but occasionally a mutation can result in physical benefits or disabilities.

Paleo-DNA (cont)

mosquitoes trapped in prehistoric amber. This is pure science fiction (for the present). According to Edward Rubin, the director of the Joint Genome Institute where the cave bear DNA sequencing was performed, DNA can only survive about 100,000 years in the conditions under which the cave bear bones were found. Dinosaur DNA would have to survive at least 65 million. (Perhaps encapsulated in amber it could.)

1. National Geographic News Special Series: The Alps, "Ancient Bear DNA Mapped – A 1st for Extinct Species", June 5, 2005, http://news.nationalgeographic.com/news/2005/06/0606_050606_alpsbears.html.
2. M. Schirber, Live Science, "Genetic Time Travel: Scientist Decode DNA of Extinct Animal", June 7, 2005, http://www.livescience.com/othernews/050607_cave_bear.html.
3. J. B. Verrengia, Live Science, "Scientist Find Prehistoric Dwarf Skeleton", October 27, 2004, http://www.livescience.com/humanbiology/prehistoric_dwarf_041027.html.

Markers

Genetic genealogy depends on matching the characteristics of certain positions on DNA called markers or *loci* (the plural of locus, Latin for place) of two or more individuals to determine if they share a common ancestor. The fewer mismatches people exhibit, the shorter the time that probably elapsed since their Most Recent Common Ancestor (MRCA).

For a DNA marker to be useful in genetic genealogy, it must be *nonrecombinant*, meaning it is inherited intact always from the mother or always from the father, giving it an unambiguous origin. Most DNA appearing in the nucleus of a cell does not qualify. Males and females inherit one member of all 23 pairs of chromosomes from the mother and the other member of a pair from the father. Members of twenty-two of the pairs form matched sets that are the same size and serve the same purpose. It is impossible to tell which member of a pair came from which parent (without a paternity test). The two members can exchange genetic material (they are recombinant) and either member can be passed on to the next generation.

Nuclear DNA

The 23rd pair of chromosomes in a female is composed of two X-chromosomes, one inherited from the mother and one from the father. Like the members of the other 22 chromosome pairs, in a female the members of the 23rd pair form a matched set, can exchange genetic material, and either member can be passed on to the next generation.

In a male, the 23rd pair of chromosomes contains an X chromosome inherited from the mother and a Y inherited from the father. Because the Y-chromosome is nonrecombinant and is handed down by a father only to his sons, it can be used in tracing the exclusively male line of a family. Since a surname is normally handed down in Western cultures through the male line, studying the Y-chromosome is a good way to study a surname. If a father has only daughters, his Y-chromosome becomes extinct. A woman cannot participate in a Y-chromosome study on her maiden name, nor can any of her children. Her sons are candidates for a surname study only on their father's family name, but her daughters are ineligible.

Mitochondrial DNA

Mitochondrial DNA is handed down by a mother to all of her children. It is useful to both males and females for tracing the exclusively female lines of their families.

Markers appearing on either the Y-chromosome or in mtDNA have unambiguous parental origins and can be used for genetic genealogy.

Another characteristic that a marker must have to be useful to a genealogist is that its mutation rate must be known to a reasonable degree of accuracy. Knowing the rate is important in providing an estimate of how close the relationship is between two people who mismatch on this marker. If a marker mutates very slowly, it is not useful for studying close relationships. If the marker mutates very quickly, even closely related people would show mismatches on the marker, so it might not be useful in determining relationships, although the marker would have value for forensic identification.

There are two kinds of markers that are used in genetic genealogy:

• Short Tandem Repeats found on the Y-chromosome (Y-STRs) mutate or change at rates compatible with the time period of written history. Y-STRs are useful for tracing exclusively male lines over the last few hundred years.

• Single Nucleotide Polymorphisms (SNPs, or *snips*) mutate very slowly and are used for long-term population studies. Since SNPs appear on both the Y-chromosome and in mtDNA, they can be used for studying deep genetic roots along both exclusively male and exclusively female lines.

SHORT TANDEM REPEATS (STRs)

STRs are the type of marker that was used by Sir Alec Jeffreys in pioneering the techniques of DNA fingerprinting. An STR

Y OH Y!

There are several aspects of the Y-chromosome that are worth mentioning.

• Y-chromosome markers are valuable in analyzing samples that contain both male and female DNA. Testing a sample for only Y-chromosome markers provides a way to isolate the male contribution to the sample. This is valuable in investigating certain types of crimes. While a full DNA profile can be used to uniquely identify an individual, the Y-chromosome can only differentiate men with different paternal lineages. Although using Y-chromosome markers will not result in an exact identification, Y-STRs are still useful in narrowing a list of suspects.

• Sperms carrying X-chromosomes are slightly heavier than those carrying Y's because they have slightly more genetic material. This allows them to be separated so that the sex of a child can be chosen during in vitro fertilization.

Continued on p. 28

Y OH Y (cont)

• There are some diseases such as hemophilia that occur almost exclusively in males. Females rarely have them but can act as carriers, transmitting the disease to the next generation.

Y This phenomenon is caused by a recessive gene that appears on the X-chromosome in both males and females in the region where the male Y-chromosome is missing a "leg."

A female can only contract the disease if she carries the hemophilia gene on both of her X-chromosomes. Since this is unlikely, a female will rarely exhibit the disease, although a woman who has it on one of her X's has a 50% chance of passing it on to her child.

Y If a male inherits the hemophilia gene on the X-chromosome from his mother, it cannot be overridden. His Y is missing a leg where that gene would appear. He will contract the disease, but he can only transmit it to his daughters, all of whom inherit his X. His sons will manifest the disease only if they inherit the hemophilia gene from their mother.

is a sequence of nucleotides that repeats a characteristic number of times. The repeating section is bracketed by two constant segments that are useful as references when determining the length of an STR. (See Figure 2.) If a sequence of DNA is thought of as the sentence, "Mary had a little lamb," if it contains an STR it might read, "Mary had a lilililittle lamb." The number of repeats is called the value for the allele of that marker, with each marker having a characteristic range of allele values (allele for short).

Y-STRs are assigned DYS numbers (D = DNA, Y = Y-chromosome, S = Segment) by an international standard body called HUGO–Human Gene Nomenclature Committee–based at University College, London.[2] For example, at the Y-STR locus **DYS391**, the base sequence may read TCTA TCTA TCTA TCTA TCTA TCTA TCTA TCTA where TCTA is repeated eight times, giving the marker an allele value of 8. Some markers, such as **Y-GATA-H4**, still carry an older non-standardized name adopted by their discoverers characterizing the repeat motif exhibited by the marker or designating its location on the genome. A person's haplotype is his set of alleles for a specified set of STR markers. Closely related indi-

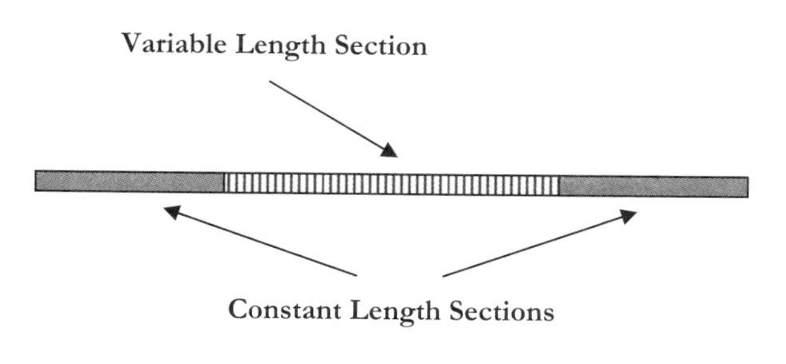

Figure 2. The structure of a Short Tandem Repeat (STR)

viduals have similar haplotypes. A typical haplotype resulting from the 12-marker test offered by one of the genealogy testing companies is shown in Table 3.

From the STR markers identified to date, a core set of loci defining the minimal haplotype for forensic applications has been designated by the Y-STR User's Group to include **DYS19** (also known as **DYS394**), **DYS385a** and **b**, **DYS389I**, **DYS389II**, **DYS390**, **DYS391**, **DYS392**, and **DYS393**. It is based on established molecular locus characteristics, the amount of worldwide population data available, locus specific mutation rates, forensic validation, and successful application to forensic case work.[3] These eight loci are those most often used in population and forensic studies[4] and are included in the list of markers used by the major genetic genealogy laboratories in their smallest test panels.

When a Y-STR marker mutates, it either adds or subtracts one or more repeats, so that its allele either increases or decreases by that amount. While most markers rarely show mutations that cause a change of more than one repeat at a time, faster mutating markers can show additions or deletions of more than one repeat occurring in a single event. A typical set of markers used for genetic genealogy contains both slower and faster moving markers as a way to derive the most information from a test panel. Slower moving markers are useful in ruling out someone as being a member of a family. This is especially valuable to groups with rare surnames where differences from a family haplotype can show up on lower power testing. Once a family group has been estab-

Table 3. Typical results of the 12-marker test.

DYS	393	390	19	391	385a	385b	426	388	439	389I	392	389II
HP1	12	23	14	10	14	17	11	17	12	13	11	29

lished by slower markers, faster markers are useful in teasing apart various family lines. In Table 4, a second haplotype has been added to Table 3. The single mismatch they exhibit is shaded.

Table 4. Comparison of two haplotypes on the 12-marker test. These two haplotypes show one single-step mismatch.

DYS	393	390	19	391	385a	385b	426	388	439	389I	392	389II
HP1	12	23	14	10	14	17	11	17	12	13	11	29
HP2	12	23	14	10	**15**	17	11	17	12	13	11	29

STRs that appear in more than one location on the Y-chromosome are called multi-copy markers. (See Figure 3). These markers are usually faster moving than single-copy markers because they contain more genetic material that can mutate. The most commonly used multi-copy markers are **DYS385a** and **b** (included in the Y-User Group's core haplotype), **DYS464**, **DYS459**, and **CDYa** and **b**. Genetic genealogy test panels normally include at least one multi-copy marker. The haplotypes shown in Table 4 show a single step mismatch on **DYS385a**.

Although the markers **DYS389I** and **DYS389II** might appear to be the components of a multi-copy marker, they are not. They are overlapping regions in DNA with **DYS389I** contained in **DYS389II**. (See Figure 4.) In calculating the values of **DYS389**, the value of the shorter component **DYS389I** must be subtracted from that of **DYS389II** to get the correct number of mismatches. For example, if one haplotype shows **DYS389I** and **II** = 13 and 29 and a second haplotype shows **DYS389I** and **II** = 14 and 30, only one mismatch is counted, since the single mutation that appears in the shorter part of the marker is also contained in **DYS389II**. If the second haplotype shows **DYS389I** and **II** = 13 and 28,

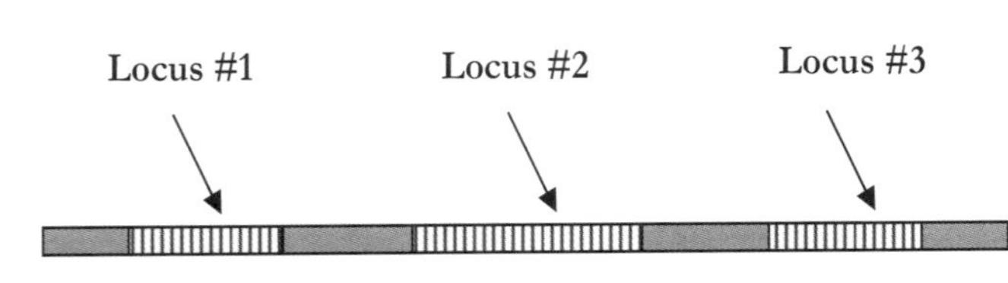

Figure 3. Sketch of a multi-copy marker.

only one mutation is present, this time located on the part of **DYS389II** that is not shared with **DYS389I**. If the second haplotype has **DYS389I** and **II** = 14 and 29 as in Table 5, two mutations are scored, one where a repeat has been added to **DYS389I**

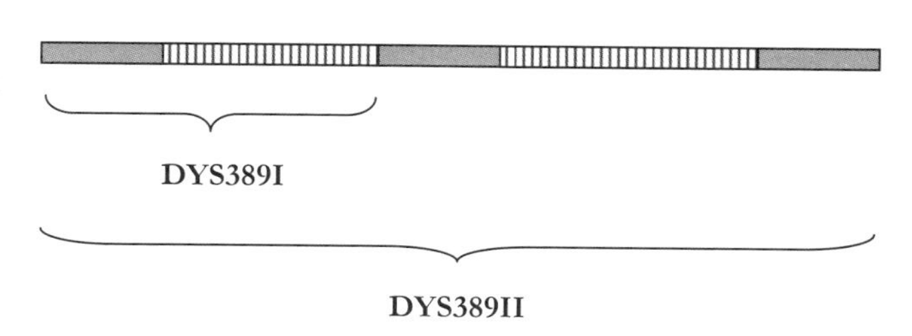

Figure 4. Overlap of DYS389I and II.

Table 5. These two haplotypes look like two mismatches but, because of the overlap of DYS389I and DYS389II, they count as only one.

DYS	393	390	19	391	385a	385b	426	388	439	389I	392	389II
HP1	12	23	14	10	14	17	11	17	12	13	11	29
HP2	12	23	14	10	14	17	11	17	12	**14**	11	**30**

and the other where a repeat has been deleted from the part of **DYS389II** not shared with **DYS389I**, causing the length of **DYS389II** to remain unchanged.

YCAII is a fast changing multi-copy marker. Its components **YCAIIa** and **YCAIIb** consist of a two-base-pair motif that is typically more volatile than longer repeat motifs. They are also more likely to increase or decrease in length by more than one repeat during a mutation. For this reason, mutations of the **YCAIIa/b** marker are counted using the infinite allele model that scores any change in length of a marker as a single mutation.[5] So no matter how large a difference between two people in a **YCAIIa** or a **YCAIIb** value, it is scored as a single mismatch.

CDYa and **b** are the fastest changing Y-STR marker used for genetic genealogy, **DYS454** is the slowest. **DYS464**, first characterized by *A. Redd et al.*[6] typically appears in four locations near the center of the Y-chromosome. They are labeled **DYS464a**,

Continued on p. 34

CHIMERAS

A DNA profile is uniquely associated with a single individual, except for identical twins. This is the basis of paternity testing and of criminal forensics. But is the reverse true? Is it possible for an individual to have more than one DNA profile?

There are rare cases in which a person has more than one DNA profile. The reason is not clear, but might be associated with the twinning process. Identical twins are assumed to derive from a single egg that divides and separates soon after being fertilized. Fraternal twins are assumed to be the result of two eggs fertilized separately. A third kind of twin can occur as the result of an egg that splits before it is fertilized, resulting in twins that are "half-identical" and "half-fraternal."[1]

If it is possible for an egg to divide to create two genetically identical people, is it possible for two eggs to coalesce to create one person? The answer appears to be yes. These people are known as *chimeras*. "Chimera" (pronounced kih-MEE-ra) is derived from the Greek name for a mythological beast that has the head of a lion, the body of a goat, and the tail of a serpent.

One type of chimerism apparently is created in the very early stages of embryonic development when both embryos are still composed of unspecialized cells[2] resulting in a chimera who has organs carrying two different DNA profiles. Occasionally a chimera has two different blood types. There are about 30 cases of such human chimerism in literature.[3]

If a chimera were to be paternity tested or have his DNA analyzed by the RFLP techniques used in criminal forensics, he would exhibit two DNA profiles. But since the Y-chromosomes and mtDNAs of his two profiles are identical, he would not show any ambiguity when his DNA was analyzed using Y-STR and Y- or mtDNA-SNP tests.

A Russian mass murderer named Andrei Chikatilo escaped identification for 12 years, partly because the DNA in his semen did not match the DNA in his blood.

Chimeras (cont)

Early in the murder investigation, he was released because the DNA of his blood did not match the DNA of the semen left at the crime scenes. Only after he was finally recaptured was it was discovered that he was a chimera.

A second type of chimera occurs when two embryos share their blood in the uterus. If the twins are not genetically identical, one or both can be a chimera with two DNA profiles in his blood. Chimeras are estimated to occur in about 7% of fraternal twins.[5] Chimerism that probably also occurs among identical siblings cannot be detected.

The British Columbia Children's Hospital reported the case of an eighteen-month-old female twin suffering from chronic lung infections. High levels of chloride in her sweat indicated she suffered from cystic fibrosis (CF), a genetic disorder. A chloride test of her twin, however, indicated no sign of the disease, although the girls were thought to be identical.

DNA tests performed on blood samples from the twins showed identical DNA, and that *neither* twin carried CF mutations. But DNA testing on skin samples revealed DNA profiles that were *not* identical, and that the sick twin carried the CF mutations. Doctors concluded that the twins developed from two embryos, implanted so close to each other on the uterine wall that their placentas merged. The blood supply of the healthy twin was used by both, so that the sick twin became a chimera. The sick twin carried her healthy sister's DNA profile in her blood and her own CF-mutated DNA in the rest of her body.[6]

1. L. Wright, *Twins and What They Tell Us About Who We Are*, John Wiley & Sons, p. 86.
2. http://www.thetech.org/genetics/ask.php?id=23.
3. http://en.wikipedia.org/wiki/Chimera_%28animal%29.
4. http://www.mayhem.net/Crime/serial1.html#chikatilo.
5. N. L. Segal, "Commentary on Ruboki RJ, McCue BJ, Duffy KJ, Shepard KL, Shepherd SL, Wisecarver JL. Natural DNA Mixtures Generated in fraternal twins *in utero*", *J. Forensic Sci.* 2001 (46)1:120-5.

DYS464b, **DYS464c**, and **DYS464d**. **DYS464** has been observed to occur as many as six times in African populations belonging to Haplogroup E.[7] (Additional copies are labeled **DYS464e**, **DYS464f**, **DYS464g**, etc.)

Because the mutation rate of **DYS464** is so high, it improves a Y-test's ability to distinguish parental lineages. In a recent study on 679 men from three U. S. populations, 179 different combinations of **DYS464a,b,c,d** marker values were found. Of these combinations, over half were found in only one person each. The most common combination, 15, 15, 17, 17, was shared by only 10.6% of the individuals tested. In comparison, testing on the next fastest mutating marker, **DYS385a,b**, produced only 56 different combinations.[8]

Because the techniques used to analyze the **DYS464** marker cannot distinguish the position on the Y-chromosome of its various components, mismatches on this marker are scored differently. The values of **DYS464** components are reported in ascending order. Because of its rapid mutation rate, **DYS464** is scored using the infinite allele model (p. 70), so that every mutation is considered as a unique event, even if it involves a marker value change greater than one. For example, if the **DYS464** maker values of two people are reported as:

DYS464	a	b	c	d
Results #1	15	15	17	17
Results #2	15	17	17	18

Since they can be rearranged as:

DYS464	a	b	c	d
Results #1	15	15	17	17
Results #2	15	18	17	17

they are only considered as having a single mutation appearing at **DYS464b**.

SICKLE CELL - THE GOOD NEWS/THE BAD NEWS

Sickle cell anemia is a genetic disorder caused by a single mutation in the gene that codes for the protein hemoglobin. Hemoglobin carries oxygen in the blood. Sickle cell anemia is present in 1 out of 500 African American births.[1] Red blood cells are normally pliable so that they can squeeze through capillaries, but sickle cells are deformed and not as flexible, causing them to obstruct capillaries and restrict blood flow. Tissues downstream from the blockage are deprived of oxygen, causing pain, organ damage, stroke, and shortened lifespan.[2]

The mutation that causes sickle cell anemia is unusual as it has both harmful and beneficial effects. The gene that carries the mutation is recessive. That is, a person will develop the disease only if the mutation occurs on both members of a pair of chromosomes. But if the mutation appears on only one member of the pair, the person has a much milder form of the disease known as sickle cell trait. He also exhibits a genetic immunity to malaria. If the mutation does not appear on

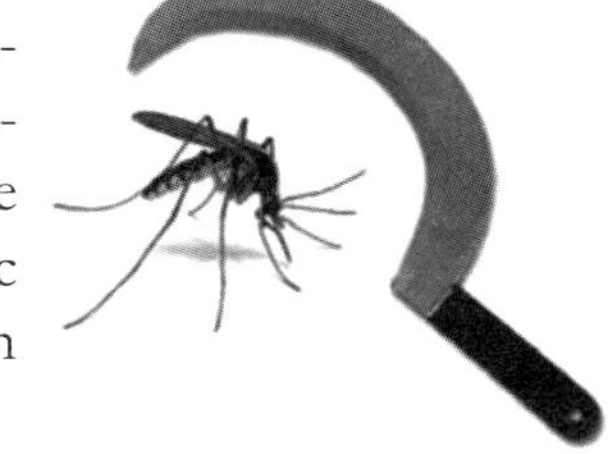

Continued on p. 36

SINGLE NUCLEOTIDE POLYMORPHISMS (SNPS)

Single Nucleotide Polymorphisms, or SNPs (commonly called *snips*) are found in both chromosomal and mtDNA. SNPs involve a base change at only one nucleotide. For a variation to be considered a SNP, it must occur in at least 1% of a population. In one part of the world such a variation may occur in less than 1% of the population while in another part of the world it may occur more than 1% of the population. Depending on where he lives, a person's DNA anomaly can be considered either a mutation or a polymorphism. SNPs, which make up about 90% of all human genetic variation, seem to occur every 100 to 300 bases along the 3-billion-base human genome. Two of every three SNPs involve the replacement of cytosine (C) with thymine (T). SNPs can occur in both coding (gene) and noncoding (junk) regions of the ge-

Sickle Cell (cont)

either chromosome, the person is not immune to malaria, nor does he develop sickle cell anemia.

If a person has two sickle cell genes, he will develop sickle cell and probably die young. If a person has two normal genes, he will not be protected against malaria, so that he might also die young if he lives in an area where malaria is common. But if he has only one sickle cell gene, he will be protected against both full blown sickle cell anemia and malaria. He will also be a carrier of the sickle cell, with a 50/50 chance of transmitting it to his offspring. Over an extended period of time half the population would acquire one recessive sickle cell gene in places in the world with a high incidence of malaria.

1. University of Maryland, Book Diseases, Sickle Cell Disease, http://www.umm.edu/blood/sickle.htm
2. http://en.wikipedia.org/wiki/Sickle-cell_anemia

nome.[9] Although SNPs do not usually cause genetic disorders, they are often found close to genes that do. SNPs can serve as biomarkers for these disease-causing genes.[10]

Since nearly all SNPs have mutated only once in human history, we know that their mutation rates are very low. A SNP is referred to as *biallelic* because it has one of two values, either the original value or the mutated value.

SNPs both on the Y-chromosome and in mtDNA are useful in population studies. By observing which groups have which SNP mutations, a timeline can be created to indicate when and where a population appeared and how it was derived from pre-existing groups. The most widely shared SNPs are regarded as the oldest, while less frequently observed SNPs are considered to characterize later populations. A group of people who share the same set of SNPs is known as a *haplogroup*. The term *clade* is more commonly used to describe a female haplogroup based on mtDNA SNPs. A diagram of haplogroups or clades showing the relationship among their characteristic mutations is called a *phylogenetic tree*.

A FEW WORDS ABOUT MUTATIONS

DNA damage, due to normal metabolic processes inside the cell nucleus alone, occurs at a rate of 50,000 to 500,000 molecular lesions per cell per day.[11] While this is only a tiny fraction of the 3 billion base pairs in the human genome, a single unrepaired lesion (called a mutation) in a gene can be a catastrophe for a cell.

Failure of DNA damage control mechanisms can result in permanent structural damage or, rarely, benefit. Mutations can arise spontaneously during cell division or through exposure to toxins in the environment. Most mutations are harmless, but mutations occurring in genes can cause a change in the production of a protein vital to cell function. This can lead to a disease such as diabetes. All types of DNA, genetic and junk DNA from the nucleus and mtDNA in the cell's mitochondria, experience mutations.

Mutations that occur in an egg or a sperm are known as *germline* mutations. STRs and SNPs on the Y-chromosome and in mtDNA fall into this category. If a germline mutation is passed to an offspring during fertilization, the mutation will appear in all cells of the offspring so that he might pass the mutation on to the next generation. This type of mutation is useful in isolating genes that are related to diseases. If several family members have a genetic disorder, their DNA profiles can be compared against family members who are unafflicted to determine where on the genome or in the mtDNA the mutation occurs. Germline mutations are also the foundation of genetic genealogy.

Mutations that are not in the germline are called *somatic* mutations. Somatic mutations do not appear in all cells of an organism, only those that are derived from the newly mutated cell. They are not passed to the next generation. Somatic mutations are common in mtDNA since mtDNA does not have the error-correcting mechanisms of nuclear DNA.

A germline mutation that appears on the Y-chromosome or in mtDNA is passed down along the exclusively male or the exclusively female line by the man or woman who first exhibits the mutation. If the man has brothers or the woman has sisters who do not have the mutation, from the next generation forward, there will be a mix of descendents, some with the mutation and some without.

Genetic genealogy can be used to sort out family lines based on STR and SNP mutations, keeping in mind it is only possible to test living people. By collecting enough samples, and observing which people with known pedigrees have mutations in common, much can be deduced about the DNA profile of the family and the genealogies of people who do not know how they are related to the family.

Figure 4 shows an example of how this works for Y-STR testing. (SNP testing on both the Y-chromosome and mtDNA would follow the same diagram, only on a much longer time scale.) **S1** and **S2** are sons of **F** who have identical haplotypes. When they are compared to other males in the family (the sons of their paternal uncles **U1** and **U2**, for example), both **S1** and **S2** match them exactly except for one mutation, indicated by the vertical hatching. Since it is highly unlikely that the sons could have experienced the same mutation independently, it is virtually certain that their father also had the mutation, but their paternal uncles did not.

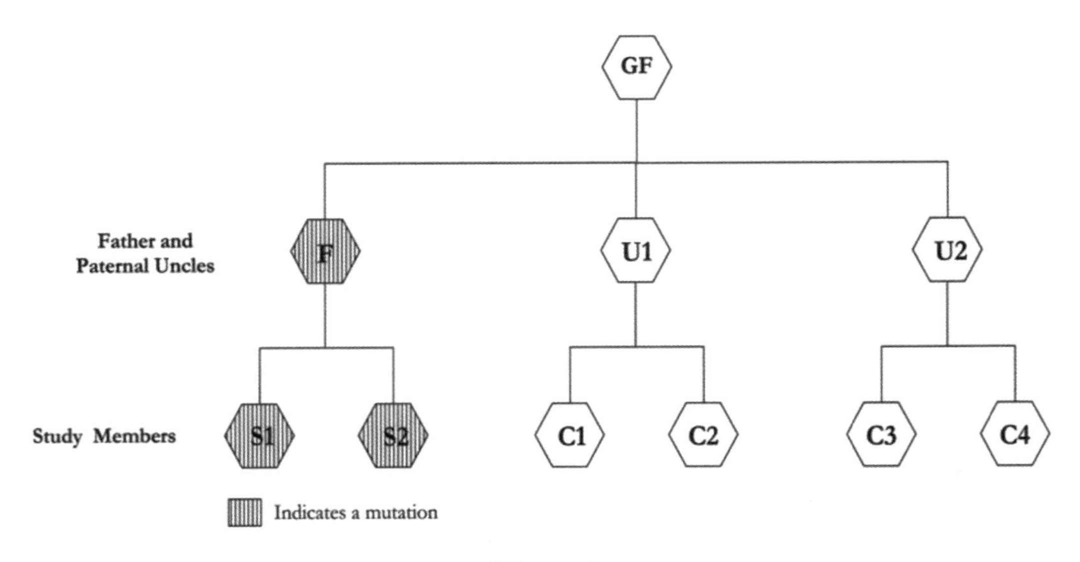

Figure 4.

In this situation the Most Recent Common Ancestor (MRCA) of the six study members is the grandfather. His haplotype, called the Modal Haplotype for the group, is derived from the most common marker values among his descendents. Since four of the six have the same haplotype, the grandfather is assumed to have had this haplotype. In a larger group, the Modal Haplotype is composed of the most frequent value of each marker, but there may be no study member who has the Modal Haplotype.

Even if a new study member is not sure of his pedigree, if he closely matches the family haplotype, he shares a common ancestor with the rest of the family. If he does not match closely, he probably does not share an ancestor in the recent past. The time when the common ancestor lived cannot be determined for certain, but can be estimated from the number of markers tested, the mutation rate, and the number of mismatches he shows to various other study members.

An excellent example of how STR testing can prove family membership is the Thomas Jefferson/Sally Hemmings study. A diagram of the results[12] is shown in Figure 5. The purpose of the study was to determine the parentage of Eston Hemmings, the youngest son of Jefferson's houseslave, Sally Hemmings. There was circumstantial evidence that he was fathered by Jefferson, but it had not been proven conclusively.

Thomas Jefferson was the only son of Peter Jefferson. Since Thomas was a bachelor with no documented children, DNA samples taken from five male descendents of Field Jefferson, Thomas' paternal uncle, were used as reference haplotypes against which to compare the other test results. Samples were also taken of a descendent of Eston Hemmings, Sally's youngest son, from a descendent of Peter Carr, Jefferson's brother-in-law, and from a descendent of Thomas Woodson. Woodson had claimed to be Jefferson's son.

Test results showed that the haplotype of Eston Hemmings' descendent matched that of the descendents of Field Jefferson. While this proved that Eston was a descendent of a Jefferson along the male line, it did not prove that he was the son of Thomas. While circumstantial evidence is strong enough to show that Eston was a product of a liaison between Thomas Jefferson and Sally Hemmings, the fact that paternity cannot be proven leaves lingering doubt for some Jefferson descendents. There were other

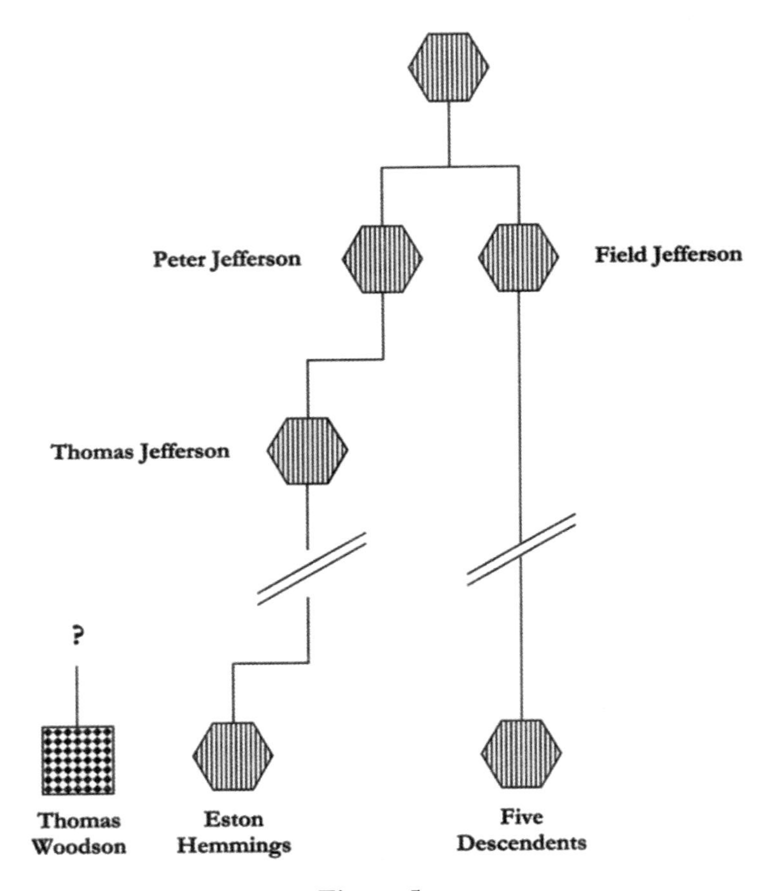

Figure 5.

Jeffersons living in the area at that time, and these groups point out that Eston could be descended from any of the others. The Woodson DNA did not match the Jefferson haplotype at all, nor did it match the haplotype of the descendent of Peter Carr, ruling out both possibilities for his paternity.

REFERENCES–THE INS AND OUTS OF GENETIC GENEALOGY

1. http://www.mumma.org/dna.htm
2. http://www.dnaheritage.com/glossary.asp
3. See the Y-STR database description at http://www.ystr.org/index_usa_gr.html
4. Y-STR Profiles in DNA, March 2003, www.promega.com
5. http://nitro.biosci.arizona.edu/ftdna/models.html#Infinite
6. Redd, A. J., A. B. Agellon, V. A. Kearney, V. A. Contreras, T. Karafet, H. Park, P. J. Kniff, M. F. Hammer, "Forensic value of 14 novel STRs on the human Y-chromosome", *Forensic Science International*, **3460** (2002) 1-15.
7. http://www.familytreedna.com/facts_genes.asp?act=show&nk=3.2
8. Butler, J. M., and R. Schoske, "Forensic Value of the Multi-copy Y-STR Marker DYS464, *Conference of the International Society of Forensic Genetics*, 2003 (poster).
9. Human Genome Project Information, SNP Fact Sheet, http://www.ornl.gov/sci/techresources/Human_Genome/faq/snps.shtml#snps
10. National Center for Biotechnology Information, A Science Primer, SNPs: Variations on a Theme, http://www.ncbi.nlm.nih.gov/About/primer/snps.html
11. http://en.wikipedia.org/wiki/DNA_repair
12. E. A. Forster et al, "Jefferson Fathered Slave's Last Child," *Nature* **396** 1998:27-28.

SURNAME STUDIES

What worked for the descendents of Eston Hemmings can work for you, too! (If you don't know who Eston Hemmings was, you're cheating and starting in the middle of this book. Go back one chapter.)

Being part of a surname study is a lot of fun. Whether you start your own and become a Group Administrator or you join an existing study, you are bound to find much interesting (perhaps surprising) information about your family. You may confirm what you already found out through paper documentation, you may discover long-lost relatives, and you might find the migrational pattern of your family. DNA testing provides you with information that cannot be obtained through conventional genealogy resources.

IS THERE ALREADY A STUDY FOR MY NAME?

First find out whether there is an existing study for your name by searching the websites of the various organizations that offer DNA testing. Even if your name is not the primary focus of a study, it might be included in the study of a related name. Therefore, make sure you search under spelling variants of your name. As of September 2005, there are about 2,500 surname projects worldwide among the different testing labs. The largest of these is Family Tree DNA, which is hosting 2,200 surname studies including 23,000 surnames and variants.

As an example of what you can do to find out if there is a study for your surname, let's use the online search facilities offered by Family Tree DNA. The Family Tree DNA page offers a search engine you can use to find out whether your name is being studied already. Start your search on the Family Tree DNA homepage under the heading "Search by Surname." Search options on the pull-down menu include: *Begins With, Equals, Contains, Ends With,* and *Sounds Like.* This last option refers to the *Soundex* code that is a four character representation based on the way a name sounds rather than the way it is spelled. Use it to identify spelling variations on a name. The search engine will calculate the Soundex equivalent for your name and return "Sounds Like" matches automatically.

For each search option, the search engine returns a list of matching names, with the number of people by that name who are being tested. The engine also gives links to the projects associated with the matching names (a name might be included in more than one study), the number of members in each study, and the first few words of each study's objectives. The list of projects might be lengthy if you have a common name, but you should be able to eliminate many as being irrelevant on the basis of nationality or geographic location.

Figure 1. Search engine results for the surname Rice using "Begins With"

The following names matched your search request:

Name	Count
Rice	147

One or more Surname Projects had an associated name that matched your request.

Project	Members	Description
Rice	154	The Edmund Rice Association...

For example, if you are searching for information on the surname Rice, the search engine returns the results in Figures 1, 2, 3, and 4 for the different search options. A "Sounds Like" search on Rice (with a Soundex value of R200) produces 58 matches

Figure 2. Search engine results for the surname Rice using "Equals"

The following names matched your search request:

Name	Count
Rice	147

One or more Surname Projects had an associated name that matched your request.

Project	Members	Description
Rice	154	The Edmund Rice Association…

Figure 3. Search engine results for the surname Rice using "Contains" (Same results as "Ends With")

The following names matched your search request:

Name	Count
Brice	3
Price	92
Rice	147

One or more Surname Projects had an associated name that matched your request.

Project	Members	Description
Bahamas DNA	62	The Bahamas DNA Project is a geographical project…
Bryson	16	A Surname Project for Brysons and related surnames…
Chilton Co. Area	6	The Central Alabama Surnames group project seeks…
Price	75	This surname project is open to…
Rice	154	The Edmund Rice Association…

Figure 4. Search engine results for the surname Rice using "Sounds Like"

The following names matched your search request:
SoundEx value R200

Name	Count
Rakow	1
Rauch	4
Rausch	1
Reece	1
Rees	4
Reese	1

[52 more names including Rex, Reyes, Rice, Rose, Ross...]

One or more Surname Projects had an associated name that matched your request.

Project	Members	Description
Chelsea, MA	12	For you who are checking this out...
Cuba	17	This area project is for people of Cuban descent...
French-Canadian	45	This project is dedicated to ...
Frosinone, Italy	75	We take...
Galicia, Spain	7	This project is open to those whose...

[28 other studies including Mexico, Rakes, Reece, Reyes, Rice, Rosoff, Ruiz....]

including Rauch, Rausch, Rauss, Rees, Reese, Reis, Rex, and Reyes. Some of these surnames do not have their own study, but are part of others. Some studies have only a few members, and some are quite large, with several hundred members. Some studies will be of immediate interest, while others will be unimportant to your family. The 58 Rice SoundEx names are associated with 33 studies focusing on related surnames and on people whose families originated in various geographical areas including Chelsea, MA; Cuba; Frosinone, Italy; and Galicia, Spain. In this case, the Rices I am interested in came from County Armagh, Ireland, so that the studies on names of Cuban, Italian, and Spanish origins are probably not relevant.

Table 5. The 20 Largest World Families Network DNA Surname Group Studies as of June 2005.

Surname	No.
Wells	360
Rose	266
Walker	260
Donald	235
Graves	218
Bolling	198
Brown	170
Meates	164
Johnson	163
Williams	160
Turner	154
Bassett	152
MacGregor	150
Hill	146
Rice	142
Barton	142
Spencer	137
Baker	136
Elliot & Border Rei	134
Bates	128

Another option for searching is the *Projects* link at the top of the Family Tree DNA homepage. The link will take you to the Surname and Geographical Projects page where the two types of projects are organized alphabetically by the first letter of the project name. There is a hotlink for each letter with the number of projects that can be found under that letter. (Rice is one of 85 surnames appearing under the letter "R".)

If you are interested in obtaining background information on the genetic genealogy community while you are searching for a surname study, you may visit the World Families Network (WFN) website at www.worldfamilies.net.

This site gives information on many items of interest, including Discussion Forums, Technical Resources, information on Surname DNA and mtDNA Testing, and on Featured Surname Projects. A calendar of events is included, with a link to the Journal of Genetic Genealogy. You can also find a list of projects with fifty or more participants. The twenty largest are listed in Table 5.

JOINING A STUDY

If you find that a study already exists that includes your surname, you will see links leading to the study's project page with a summary of what the project is about and its status. The page contains information on its objectives, the number of participants, and either a project web address or contact information for the group administrator. If required by the Group Administrator, there is also a Project Join Request Form that will ask for your contact information and your reason for joining the study. This information will be relayed to the Group Administrator who will contact you about joining the project.

If your name is very common, e.g. Smith, Johnson, or Jones, it may be included in several studies that each focus on a particular branch of a family or a geographical location. However, the Group Administrators for studies that share a surname are usually in touch with each other. If you contact an administrator, he will discuss the objectives of the study and put you in contact with another administrator if it is appropriate for you to join a different study, considering your family background.

The Group Administrator can also fill you in on the practical aspects of joining the group. One important decision you should discuss with him is how many markers you should get tested on. The answer to this question depends on a number of factors, including how common your surname is. If it is rare, testing on only a few markers can be enough to show whether you have a genetic connection to the family.

You can also find out how mature your name study is, that is, how many members there are and how many markers they have been tested on. A study on a common name can show many genetically unrelated groups. If you are only interested in knowing which of these groups you belong to without knowing any further details, testing on a few markers may be adequate.

Another consideration is cost. If you want to be conservative, you may get tested on the fewest number of markers and upgrade later at a slightly higher overall expense. (This might be the best option for a rare surname anyway.) You might also ask the Group Administrator whether the group has established a sponsorship program. Family Tree DNA has established a procedure for projects to set up a general fund for sponsoring new members. The amount taken from the project fund is up to the Group Administrator and can cover all or part of a test. If financing the test is an issue, you should discuss with your administrator whether you might qualify for sponsorship.

REFERENCES–SURNAME STUDIES

1. To join, send a blank email to Genealogy-DNA-L-request@rootsweb.com with the single word, Subscribe, in the subject box. To unsubscribe, do the same thing, except use the word, Unsubscribe. Send messages to the list using the address Genealogy-DNA-L@rootsweb.com.
2. http://www.familytreedna.com/forum/
3. http://www.wfnforum.net/index.php
4. http://genforum.genealogy.com/
5. Lauren Boyd, "How to Win Friends and Influence List Administrators", May 5, 2003, http://archiver.rootsweb.com/th/read/GENEALOGY-DNA/2003-05/1052165070

WHAT KIND OF TEST IS RIGHT FOR ME?

What kind of test you should take to get involved in a surname study depends on what you want to get out of being tested.

There are several different kinds of testing available through the major DNA testing companies that specialize in genetic genealogy. Y-STR testing is the kind of testing done by surname study groups that produces results relevant to genealogical time periods of less than about 1,000 years. Y-SNPs are more slowly changing markers that indicate deep ancestral roots along the exclusively male line on the order of a few thousand years or more. Y-SNP testing can indicate which haplogroup a man belongs to. STR and Y-SNP testing can only be done on males and can only provide information on their direct male lines. For example, a man cannot be STR or Y-SNP tested for his maternal grandfather's family since there is a woman (his mother) in this line of his family.

Testing on mtDNA SNPs can indicate deep ancestral roots along the exclusively female line. It can tell you which female haplogroup you belong to, commonly called a clade. Both men and women can be mtDNA-tested, but it can only provide information along their direct female lines.

SNP testing can indicate your roots only along your exclusively male (Y-SNP) or exclusively female (mtDNA SNP) family lines. Many people are certain that they have either Native American or African American ancestry, but are surprised when SNP testing indicates European roots. This result comes about because their direct male or female lines have a European origin.

MAJOR TESTING COMPANIES

The three main companies that specialize in genetic genealogy are Family Tree DNA, Relative Genetics, and DNA Heritage. Family Tree DNA and Relative Genetics offer a variety of products for Y-chromosome and mtDNA testing. DNA Heritage only offers Y-testing. A summary of the testing options offered by the three major companies can be found in Table 1. More information can be found on their websites.

There are several other companies that specialize in DNA testing for genealogy. For example, Oxford Ancestry was founded by Dr. Bryan Sykes in 2001 in response to the success of his book *The Seven Daughters of Eve*. The company originally offered only mtDNA testing, but in 2003, Oxford Ancestry expanded its product line to include haplogroup typing services based on Y-chromosome STR analysis. For a complete list of DNA testing companies, see http://www.duerinck.com/dnalabs.html.[1]

Family Tree DNA (www.familytreedna.com or www.ftdna.com)

Family Tree DNA was founded in 1999 by Bennett Greenspan and Max Blankfield in conjunction with Dr. Mike Hammer, Director of the Genomic Analysis and Technology Core testing facility at the University of Arizona, Tucson, AZ. Dr. Hammer's laboratory offers multi-user state-of-the-art facilities for the development of new biotechnologies, and meets the stringent peer-review requirements of the biotechnology research community. Dr. Hammer, along with Dr. Alan Redd, was responsible for the discovery of many of the markers that are commonly used by testing companies for both forensic and genealogical STR analysis.[2]

Relative Genetics (www.relativegenetics.com)

Relative Genetics, located in Salt Lake City, serves as the business unit of Sorenson Genomics, the DNA processing laboratory founded by the Sorenson Molecular

Table 1. Products offered by the major genetic genealogy testing companies

	Relative Genetics			Family Tree DNA			DNA Heritage		
Y-chromosome Analysis: Markers/ Alleles	15/17	24/26	37/43	10/12	19/25	31/37	23 alleles	Each add'l allele	43 allele pane

	Family Tree DNA	Relative Genetics
Native American Patrilineage Testing	Estimated from STR results. SNP confirmation tests available	Based on 3 SNP markers. Results also compared to STR database of Native Americans haplotypes
Native American Matrilineage Testing	Based on mtDNA HVR1 region + 10 SNPs. Price depends on level of test refinement	Based on HVR1 region of mtDNA
African American Patrilineage Testing	Estimated from STR results. SNP confirmation tests available	Based on 9 SNP markers. Haplogroup derived from large database of Western African populations
African American Matrilineage Testing	Based on mtDNA HVR1 region + 10 SNPs. Price depends on level of test refinement	Based on mtDNA HVR1 region. Haplogroup derived from database
mtDNA //full Sequence	Complete mitochondrial sequence	N/A

OETZI

On September 19, 1991, two hikers in the Tyrolean Alps found the frozen body of a man on a rock shelf at an altitude of 10,500 feet. He was at first believed to be the victim of a recent homicide, but it was soon discovered that the remains were that of a hunter who had died between 5,300 and 5,100 years ago. His body had been protected for millennia by the snow and ice in the mountains, and only became exposed when unusually warm weather caused his ice tomb to melt.

The Ice Man, (nicknamed Oetzi for the Oetz Valley near where he was discovered) was found along with the clothes, tools, weapons, and even medicinal herbs he was carrying when he died. His hat lay a few feet from his head where it had fallen as he collapsed. Analysis of the radioactive isotopes in his teeth and bones indicated that Oetzi was about 46 years old when he died (old for those times), and located his place of birth to within a few valleys about 40 miles southeast of the discovery site.[1] Trace amounts of arsenic and copper found in Oetzi's body indicated that he was a coppersmith by trade.

Continued on p. 55

Genealogy Foundation (www.smgf.org).[3] SMGF is a nonprofit organization that offers a free on-line database of Y chromosome haplotypes linked to surnames, dates, and places of birth.[4]

DNA Heritage (http://www.dnaheritage.com)

DNA Heritage was founded in 2002 to service the growing genetic genealogy market in Europe. In August 2004 DNAH opened a North American office in Rochester, New York. DNA Heritage has made an agreement with Relative Genetics to use their testing laboratories. DNA Heritage specializes in Y-chromosome testing only. DNA Heritage offers the flexibility of a per-marker price for testing and allows the customer a choice of which will be tested, with a minimum of 23 alleles. The company does not sell lower resolution Y-chromosome tests. DNA Heritage tests all markers at one time; when an upgrade is ordered, the results are delivered immediately.

TESTING OPTIONS

All three major companies offer STR testing to determine a man's haplotype. His results are compared with others in his surname study. The number of mismatches he has with other members of the study is an indication of how closely he is related, if at all. More specific information on how the number of mismatches can indicate the closeness of a relationship is discussed in the chapter *Are We Really Cousins?*

Family Tree DNA and Relative Genetics also perform testing on SNP markers on both the Y-chromosome and mtDNA. Because of the slow mutation rates of SNP markers, SNP testing is not useful for determining family relationships. It can however be used to determine which haplogroup someone belongs to. Y-chromosome SNP testing can only be done on males, and indicates haplogroup membership along the exclusively male line of a family. mtDNA SNP testing can be done on both males and females and indicates haplogroup (clade) membership along the exclusively female line of a family. Note that Y-chromosome and mtDNA SNP testing is not valid for tracing mixed-gender family lines, for example, through the maternal grandfather or paternal grandmother. It is useful for tracing the heritage of only the exclusively male or exclusively female lines of a family.

Because of the high cost of SNP testing, Family Tree DNA provides with its

Oetzi (cont)

The Iceman's belongings included seven articles of clothing and 20 different items of daily use. These artifacts were made of 17 types of wood and plant material, used for tools, weapons, containers and fire-making. His belt held up a leather loincloth, and leggings made of animal skin had been attached to it by suspended leather strips serving as garters. Oetzi wore a jacket, possibly sleeveless, made from alternating strips of different colored deerskin beneath an outer cape of woven grasses or reeds. A cap made from bear fur had been fastened below his chin with a strap. He wore much-repaired shoes of calf skin insulated with grasses to protect his feet from the cold.[2] There were over 30 different types of mosses and liverworts clinging to Oetzi's clothing.[3]

Extensive DNA analysis of the contents of Oetzi's intestines, of the plants he used for his clothing, tools, and herbs, and even of Oetzi himself, have been combined with the results of anthropological and paleobiological analysis to yield a fascinating window into the daily life of the Late Neolithic period. They have also helped trace Oetzi's

Continued on p. 56

Oetzi (cont)

whereabouts during the eight hours before his death 5,300 years ago. DNA analysis of blood on his knife, his arrows, and his clothes revealed much about the circumstances of his death.

DNA analysis of the contents of Oetzi's intestines revealed the composition of his last two meals and where and when he had eaten them. Pollen he accidently consumed clinging to his second to last meal showed that it had been eaten in a nearby mid-altitude conifer forest with hazel, birch, and hop hornbeam trees. It also revealed what time of year Oetzi had died, as the hop hornbeam tree blooms between March and June.

DNA taken from muscle and bone fragments found in Oetzi's intestines were identified through comparison with the genomes of the ibex and red deer. Oetzi had eaten a meal of ibex, followed by a final meal of red deer. Both meals were accompanied by unleavened bread made of einkorn wheat, one of the few domesticated grains used in the Iceman's part of the world at that time.[5]

Continued on p. 58

Y-STR results an estimate of which haplogroup an individual belongs to, based on a comparison of the results with its database of people who have been tested on both STR and SNP markers. An SNP test to verify its estimate of the most probable haplogroup is available at low additional cost. If the initial SNP test gives negative results–that is, the person does not belong to the predicted haplogroup–Family Tree DNA will continue, at no further cost to the customer, to test closely related SNPs until a match is made or until it has tested all the SNPs within its capabilities. The average number of additional SNP tests required is about three, but in rare cases, it can be as many as 15.[5]

Y-STR Tests for Surname Study Groups

If you are considering joining a surname group, you will be concerned with choosing among the available Y-DNA STR testing options. If you want to be conservative, it's possible to start small and to upgrade later. Usually there is enough DNA in a sample for an upgrade so that you will probably not be required to submit another kit if you want to test on more markers later. There is a lot of information in the chapter *How Many*

Markers? How Many People? about what you can learn from the different levels of testing. Much depends on the type of surname you have.

For example, Family Tree DNA offer three levels of Y-DNA STR testing. (See Table 2 for a list of the markers included in each test.) Each panel is composed of both quickly and slowly moving markers. A match on the slower markers tells you more about the deep family roots, the faster moving markers help define which family line you belong to. The more markers you test on, the more accurately you will be able to approximate the time to the Most Recent Common Ancestor (MRCA) you share with others.

For example, the **Y-DNA12** test offered by Family Tree DNA includes the core set of markers as defined by the Y-STR User's group, plus **DYS388**. A twelve marker test is valuable for ruling out a genetic connection between two people. Studies on rare surnames also find that the 12 marker test is sufficient to determine who does or does not have a relationship to the core group of the family.

Upgrading to more markers from a 12 marker test is desirable to differentiate

Table 2. Y-DNA test markers

Y-DNA12 MARKERS	Y-DNA25 MARKERS = Y-DNA12 + THESE 13 MARKERS	Y-DNA37 MARKERS = Y-DNA25 + THESE 12 MARKERS
DYS#	DYS#	DYS#
393	458	460
390	459a	GATA H4
19*	459b	YCAIIa
391	455	YCAIIb
385a	454	456
385b	447	607
426	437	576
388	448	570
439	449	CDYa
389I	464a	CDYb
392	464b	442
389II	464c	438
	464d	

*Also called DYS394

Oetzi (cont)

It was first believed that Oetzi was a shepherd who had been caught off guard by a snowstorm. But in 2001, ten years after he was found, a radiologist discovered an arrowhead embedded in Oetzi's left shoulder near the carotid artery. Its position corresponded to a small tear in the grass cloak that Oetzi had been wearing when he died. What had been an academic research project became a murder investigation.

Further analysis showed bruises on Oetzi's chest and defensive wounds on his hands. DNA of four different people was found on Oetzi's personal belongings: two human blood signatures on one of his arrows, one on his knife, and one on his cloak.

The evidence indicates a scenario in which Oetzi had killed two people with the arrow, retrieving it each time. The arrow shaft shattered when he missed the third shot. He was in the process of mending it when he was killed.

DNA analysis of the blood on his cloak indicated that he had been carrying a wounded companion (or he was being

Continued on p. 59

among closely related family lines. This might be a good idea, for example, if your name is rare and the people in your study descend from several brothers who came to America during colonial times.

For surnames that are not rare, a 12 marker test is sufficient to rule out a relationship, but testing on 20 or more markers is required to establish to a high level of confidence that two people are related. The **Y-DNA25** marker test is appropriate in this case. While testing on a higher number of markers will likely produce a larger number of mismatches, the calculation of the MRCA expects more mismatches as more markers are tested. For two people who show more mismatches on the **Y-DNA25** test than on the **Y-DNA12** test, the 25 marker test may indicate a closer relationship than did the 12 marker test. The closeness of the match also depends on which markers are mismatched. When tested on the same set of markers, two individuals who mismatch on one of the more slowly moving markers will probably be more distantly related than two who mismatch on a fast moving marker.

The **Y-DNA37** test can tease apart closely related family lines. It is useful to people who have common names and who have established to a high probability that they are related. It may also distinguish between otherwise genetically identical family lines. The number of mismatches, the number of markers, and which markers show a mismatch are all important in interpreting 37-marker results in order to determine the time to the MRCA.

Family Tree DNA offers upgrades to more markers through their **Y-Refine 12to25**, **Y-Refine 25to37** and **Y-Refine 12to37** testing options. They also offer the **Oxford Conversion Kit**, **Oxford Conversion Kit Plus**, and the **Ancestry Conversion Kit** to test markers not included by other testing companies so that your results are compatible with those obtained at Family Tree DNA. This also adds you to the Family Tree DNA database and allows you to submit your results to the National Geographic Genographic project, for which Family Tree DNA is doing the testing in the U.S.

SNP Estimates and Testing Options

Although membership in a haplogroup is based on which SNPs a person has, it can be estimated from a set of STR marker values. Because SNP testing can be expensive, Family Tree DNA provides an estimate of which haplogroup an individual belongs to based on a comparison of his Y-STR results

Oetzi (cont)

carried) shortly before he died. The arrow's entrance wound in his back was such that he could not have pulled the shaft out himself. Oetzi was not alone when he was injured.[6] It was well known by hunters of the Neolithic era that the best way to kill an animal was by shooting it in the left shoulder. Oetzi evidently died at the hands of a hunter.

Although Oetzi's nuclear DNA was too degraded to be analyzed, there was sufficient mtDNA to identify mutations at **16224C** and **16311C**, indicating that he was a member of haplogroup **K**.[7] (The identification was done on the basis of only ten genome equivalents per gram of tissue.[8]) This haplogroup dates to approximately 16,000 years ago, and it has been suggested that individuals with this haplogroup took part in the pre-Neolithic expansion following the Last Glacial Maximum.[9] It is most closely related to mitochondrial types in central and northern European populations.[10]

Continued on p. 60

Oetzi (cont)

1. W. Mueller, H. Fricke, A. N. Halliday, M. T. McCulloch, J-A Wartho, "Origin and Migration of the Alpine Iceman," *Science*, **302**, p 862 - 866.
2. http://www.tribuneindia.com/2002/20020303/spectrum/main4.htm
3. http://www.gla.ac.uk/ibls/DEEB/jd/otzi.htm
4. F. Rollo, M. Ubaldi, L. Urmini,I. Marota, "Oetzi's last meals: DNA analysis of the intestinal contents of the Neolithic glacier mummy from the Alps," *Proc. of the Nat. Academy of Science*, **99** (20):12594–99.
5. http://www.angelfire.com/me/ij/oetzie-lastmeal.html
6. http://www.mummytombs.com/mummylocator/featured/otzi.news.htm
7. http://en.wikipedia.org/wiki/%C3%96tzi_the_Iceman
8. O. Handt, *et al*, "Molecular genetic analyses of the Tyrolean Ice Man," *Science*, **264**(5166):1775-8.
9. http://www.kerchner.com/haplogroups-mtdna.htm
10. O. Handt, *op cit*.

with the database of the Genomic Analysis and Testing Core Facility at the University of Arizona. Knowing your male haplogroup will tell you if you have Native American, African American, or Scandinavian ancestry along your exclusively male line. An SNP test to verify the estimate of the most probable haplogroup is available at a moderate additional cost.

Note that the **Y-DNA Universal Male Test** that Family Tree DNA offers for determining your haplogroup is the same as their **Y-DNA12** STR testing option.

mtDNA Testing Options

Testing on mtDNA can determine your deep genetic roots along your exclusively female line. Family Tree DNA offers three mtDNA tests for this purpose. Because mtDNA is inherited from the mother by both men and women, the test can be taken by either males or females.

Family Tree DNA's **mtDNA** test examines 569 base pairs, including the entire **HVR1** region of the mtDNA genome. Because of the very slow mutation rate of this mtDNA, the test indicates your deep ethnic and geographic origins along your direct maternal line by identifying your female haplogroup (commonly called a clade). This test indicates if your exclusively female line is Native American in origin and if so, which of the five major groups that settled in the Americas you most likely descended from, but it does not tell you which tribe you descend from. It can point to African Ancestry or other ethnic origins along your exclusively female line.

The **mtDNA Plus** test includes the mtDNA sequences tested in the **mtDNA** test, and an additional 574 base pairs that include the entire **HVR2** region of the mtDNA genome. It examines a total of 1143 base pairs to provide an indication of your clade membership to a higher resolution. The **mtDNAPlus** test does not provide a high resolution indication of a person's clade membership unless the clade happens to have a defining mutation in the HVR-2 region. The clade assignment listed on the personal page will not change.

The mtDNA Refine test is offered to customers who have already taken the **mtDNA** test and wish to upgrade their testing to include the **HVR2** region.

The **mtDNA Full Sequence** test is offered by Family Tree DNA tests the entire mtDNA genome. The purposes of this test are a) to provide the only (or final) mtDNA test that anyone will ever need to take; b) to more accurately establish the actual mutation rate of the female inherited mtDNA; c) to apply statistics to more accurately predict when two people who match identically likely shared a MRCA; and d) to allow DNA fanatics to have results that will be 'in hand' when more advanced scientific papers are published.

Combined Testing Options

It is possible to purchase any combination of Y-DNA and mtDNA tests from both Family Tree DNA and Relative Genetics.

WHAT TO EXPECT WHEN YOU GET TESTED

Doing a DNA test these days is a piece of cake. Because the PCR techniques used have an amplification step in which the DNA in a sample is multiplied, it can analyze very small amounts of genetic material. A DNA test involves swiping the inside of your mouth with a cotton swab to collect cells from the inside of your cheek. A blood sample is not necessary.

When you sign up to get your DNA tested, you will receive a personal web page where you can track the status of your sample and view your results when they are ready. It is for your eyes only, except that your Group Administrator will have access to it if you are part of a surname study. Your personal page will provide you with information on matches found in the company's database. You will find the names and contact information for individuals you match who have agreed to release this information. In addition, Family Tree DNA provides information on your recent ethnic origins, and on the haplogroup predicted from your STR results.

You can set your preferences regarding whether you want to share (or view) names and email addresses with people who match you. You can also choose which level matches you want displayed on your page. If your name is very common, you may find many exact matches on the 12-marker test with individuals who are not closely related. You may want only matches on the 25- or 37-marker tests displayed.

In a few days after you order, you will receive a test kit in the mail with instructions on how to take a DNA sample. There will be two or three cheek swabs in the kit with instructions on how to collect your sample. A kit also comes with a release form that allows the company you are testing with to share your name and contact information with someone else who is a match, if you so desire.

Your results will be ready in several weeks and you will be notified that they are posted on your personal page. While you are waiting for your results, you may want to read background information on the testing procedure and to learn what kinds of information you might obtain from your results. It might also be a good idea to make contact with other members of your study and compare written genealogies. There is a library section on every company's website for you to use, in case you are interested in more in-depth reading.

With your test results, you are notified about close matches. If you have signed a release form, your contact information will be given to the person you match, otherwise no one will be able to see or compare with your results.

If you take a Y-STR test with Family Tree DNA, you will receive a certificate and a report generally describing Y-DNA sequencing and the meaning of haplotype matches. You will also receive an estimate of which haplogroup you belong to based on your STR profile.

If you take an mtDNA test with Family Tree DNA, you will receive a report generally describing mtDNA, the meaning of polymorphism, your differences from the Cambridge Reference Sequence and the meaning of mtDNA matches.

REFERENCES–WHAT KIND OF TEST IS RIGHT FOR ME?

1. http://www.duerinck.com/dnalabs.html
2. See for example, Alan J. Redd et al, "Forensic value of 14 novel STRs on the human Y-chromosome," *Forensic Science International*, **130**(2-3):97-111,(2002).
3. www.relativegenetics.com. Click on "About Relative Genetics".
4. www.smgf.org.
5. Bennett Greenspan, private communication.

ARE WE REALLY COUSINS?

Genealogists and non-genealogists alike join DNA studies to connect to other family members to find their roots. Whether you believe that you descend from early immigrants who arrived in America on the Mayflower in the 1600s, or you know you are the grandchild of an untraceable Russian immigrant who changed his name upon entry into the U.S. after World War II, like all human beings, you have a need to know who you are and where your family came from.

DNA can help you determine your family origins in the recent past–over the last few hundred years, and in the more distant past–over the last few thousand, depending on what tests you take. DNA can connect you with family members along your exclusively male line through Y-STR testing if you share a common ancestor over the last few hundred years. Beyond that DNA can connect you along either the exclusively male or the exclusively female lines though Y- or mtDNA-SNP tests. To research mixed male and female ancestry you must rely on paper documentation, or test the appropriate relative, if available. The DNA handed down along mixed family lines containing both male and female ancestors is more difficult to analyze, as the offspring in each generation inherit an unpredictable mix of DNA from each parent. The nonrecombinant parts of DNA, that is, the Y-chromosome and mtDNA, are not handed down through mixed gender lines.

Online databases are becoming increasingly important for the purpose of cross-matching related people with different family names. No longer is it sufficient to check out only those people with your own name. Your ethnic background and the geographical location of your ancestors affects the chances of finding matches.

As the number of people increases who have had their Y-DNA tested, so does your chance of finding that you are related to someone with a different family name. This can happen for a variety of reasons including adoption, illegitimacy and a name change. This may be the case for a surname adopted by an immigrant who was a political refugee afraid to disclose his original name to even his closest family members. Or you might match someone with a different surname because two closely related ancestors chose different family names.

The purpose of some studies is to find matches among people with the same surname descended from a specific ancestor. This is the focus of a number of group studies researching the family lines of immigrants to the U.S. in the 1600s. Others search within surname variations for a match. Finding surname matches is also becoming increasingly important to adopted children searching for their birth parents.

THE MOST RECENT COMMON ANCESTOR

Once you find a DNA match with someone else, the next questions are, "How are we related? How long ago did our Most Recent Common Ancestor (MRCA) live?"

The MRCA is exactly what the name implies: the most recent ancestor shared by two or more people. In the case of a group, the most probable haplotype of the MRCA, called the group's modal haplotype, can be constructed from the most common value of each marker. This is assumed since the unmutated value of each marker has existed longer than any mutation hence we expect more descendants to share it. The assumed haplotype of the MRCA (the modal haplotype) may not be that of anyone in the study group. In Table 1 in the chapter *How Many Markers? How Many People?* we gave the example of the Villareal study in which no member of the group had its exact modal haplotype.

Because the mutation rate of a Y-STR marker is relatively fast, Y-STR testing can be used to indicate if two people have a common ancestor in the last few hundred years. Mutation rates are statistical information, so that an MRCA calculation can only provide the *probability* that a common ancestor lived within a given number of generations or years. Its calculation is based on how many markers are tested, how many mismatches are observed between the people, and the mutation rates of the mismatching markers. Y-STR testing cannot tell you specific details of the relationship.

Since the mutation rate of SNPs is very low (0.33 per marker per million years for SNPs in the Hypervariable Region 1 (**HVR1**) of the mtDNA genome)[1] they are not useful for predicting common ancestors in the last few hundred years. SNPs, both on the Y-chromosome and in mtDNA, are used to determine deep ancestral roots over thousands of years.

MANUSCRIPTS & MUTATIONS

Scholars of ancient texts use many of the same techniques in interpreting the history of manuscripts as genetic genealogists do in interpreting DNA results. The investigation of which manuscripts descend from the same original and what that original version must have said is equivalent to a genealogist's determining which people with a surname have a common ancestor and what the haplotype of that ancestor probably was.

Until the invention of the printing press in the 15th century, new copies of manuscripts were carefully produced by hand by scribes. However careful a scribe was, he inevitably made small errors in transcription that would appear in the new manuscript. This manuscript would in turn be copied by another scribe who would try to be as accurate as possible in copying the previous scribe's transcription, unknowingly including the previous scribe's errors with a few new ones of his own.

In examining a group of transcriptions of the same material, we can learn much about the origins of the manuscripts by looking at the errors. If they

Continued on p. 65

Manuscripts (cont)

are nearly identical, we assume that they had a common ancestor manuscript few transcriptions previously. If they mismatch in a few places, they have a common ancestor, but there have been several transcriptions in the interim. Manuscripts of the same material with many differences are probably not descended from the same original.

The analogy between paleography and genetic genealogy goes further. The number of copies of a text probably increased rapidly at first, then gradually leveled out as it approached the maximum number of copies that the environment could sustain. That would be the number of libraries and individuals who wanted a copy.

The world of books also has its own versions of invasive species. For instance, in 12th-century Spain, early medieval geometry textbooks quickly became obsolete after advanced Arabic volumes arrived on the scene.[1]

1. Erica Klarreich, "Manuscripts as fossils: population-biology equations estimate medieval texts' likelihood of survival," Science News, April 9, 2005.

MISMATCHES

The fewer mismatches you have with someone else for a given number of markers tested, the more recently your common ancestor probably lived. Is it possible to match someone exactly but have an MRCA who lived many generations ago? Sure, but probably not, because if this were true there would have been enough time for a mutation or two to occur. Is it possible that one or both of you had a marker that changed and then changed back to its original value, pushing the date of your MRCA further into the past? Sure, but very unlikely that the same marker experienced two mutations that canceled each other out while not mutating on any other marker.

The greater the number of mismatched markers, the more distantly related you are. Is it possible that you mismatch someone on a large number of markers and have an MRCA who lived only two or three generations ago? Sure, but it is unlikely that so many mutations could have occurred in a short period of time. If you mismatch on more than a few out of twenty markers, you are not considered to be related in a genealogical time frame.

Different Y-STR markers have different mutation rates. A typical test panel includes both slow and fast changing markers. The slower markers are useful for determining whether two people belong to the same family group, while the faster changing markers indicate more about the details of how they are related within the family group. Markers with shorter repeat motifs (such as the **YCAIIa/b** marker with a repeat containing only two nucleotides) tend to have high mutation rates. Because there is more genetic material in multi-copy markers (**DYS385**, **YCAII**, **DYS464**, **DYS459**, and **CDY**), they also tend to have high mutation rates.

The composition of a test panel is also important in determining how accurate an MRCA estimate is. A panel should include markers with a wide range of mutation rates. The accuracy of an MRCA calculation based on individual mutation rates is greater than the accuracy of a calculation based on the average rate even when using a large number of markers. For example, Family Tree DNA incorporates marker-specific mutation rates in its online Family Tree DNATiP (Family Tree DNA Time Predictor) estimator created by Dr. Bruce Walsh at the University of Arizona. These mutation rates were obtained in collaboration with the GATC (Hammer) Laboratory at the University of Arizona and were first announced by Family Tree DNA 1st International Conference on Genetic Genealogy in Houston on Oct. 30, 2004. The power of an MRCA calculation based on the Family Tree DNA 37-marker panel using the estimator is similar to a 56-marker test using an average mutation rate of 0.004, or 110-marker test based on an average rate of 0.002.

GENETIC DISTANCE

The genetic distance between two people is an indication of how far apart they are genetically. Ideally, the genetic distance would equal the number of mismatches. But this makes the assumption that all markers have the same mutation rate, which is not true. Slowly changing markers seem to change in a stepwise fashion, that is, adding or subtracting one unit of length during each mutation. More rapidly changing markers seem sometimes to change several steps at a time, so that no matter how many mismatches they show, the difference should count for only one mutation. It is the challenge of genetic genealogists to translate the observed number of differences between two people into the actual number of mutations that have occurred and their frequencies to arrive at the most accurate genetic distance.

The **stepwise model** of mutations assumes that a single mutation changes the length of a marker only one unit of length. The stepwise model says that a (+1) mutation, where the allele increases by one, has the same probability as a (-1) mutation, where the allele decreases by one. The stepwise model allows for multiple successive mutations on a single marker, including mutations that cancel each other.

The stepwise model also includes the possibility that two separate family lines experience the same mutations on the same markers, maintaining a perfect match on these markers. Since a mutation could have reversed itself, the stepwise model gives a somewhat higher estimate for the number of generations (or years) to an MRCA than does the infinite allele model.

This model assumes that all mutations are single-step, so that a difference of more than one on a marker is counted as that number of mutations. The genetic distance based just on these stepwise markers is equal to the number of mutations. If a marker shows a two-step mismatch, it assumed to have occurred in two separate events. For example, if one person's haplotype has **DYS439** = 10 and another person's has **DYS439** = 12, they are considered to have two single-step mutations, for a genetic distance on this marker of 2.

The **infinite allele model** assumes that the possible outcomes of a mutation are infinite, that is, they do not occur through only single unit changes in length. This means that there is a negligible chance that two different markers will mutate in the same way, so that each mutation is unique. Using this model, the only way two individuals can match on a marker is for the marker not to have changed. The model does not take into account the possibility of two or more single-step mutations on the same marker, including those where a mutation might have canceled in a two-step process, changing in one direction the first time and then in the reverse direction the second time.

The infinite allele counts a mutation on a marker as a single event, even if the marker changes more than one unit in length. Using the earlier example, if one person's haplotype has **DYS439** = 10 and another person's has **DYS439** = 12, according to the infinite allele model, they are considered to have only one mutation, for a genetic distance of one.

The calculated genetic distance is less, and therefore the probability of finding the MRCA within a given number of generations will be higher using the infinite allele method than that calculated using the stepwise model. In other words, for the same mismatches on the same markers, the infinite allele model predicts an MRCA shared by two people will not be as far in the past as predicted by the stepwise model.

In its calculation of genetic distance, Family Tree DNA takes into account which markers should be scored according to which mutation model. If you show a multi-step mismatch with someone on a marker, the calculation of your genetic distance will depend on which marker has experienced the mutation(s).

MRCA PROBABILITIES—THE FTDNATiP™ CALCULATOR

Once you know your genetic distance from someone else, the next step is to calculate when your MRCA probably lived.

DOWN SYNDROME

Down Syndrome (DS) is one of the more common genetic birth defects and is one of the most common reasons for mental retardation. About 1 in 900 people is born with this disorder. Because of the higher fertility rates in younger women, the majority of DS children are born to mothers under 35 years of age. However, the likelihood of having a DS child increases with advancing maternal age.

People with DS have an excess of genetic material in their cells–specifically too many chromosome 21s. The majority, approximately 95%, have a complete extra chromosome 21. Four percent have a fragment of chromosome 21 attached to another chromosome, usually chromosome 14. About 1% have a mixed, or mosaic, pattern of cells in their body with some normal cells and other cells with an extra chromosome 21.

Most cases of DS result from faulty sperm or egg cell (gamete) division and are not inherited. But a small percentage of those with Down inherited the disorder from a parent.[1]

1. http://www.ygyh.org/ds/whatisit.htm

The FTDNATiP™ (Family Tree DNA Time Predictor) online calculator estimates the probability of finding the MRCA shared by two people, taking into account both their genetic distance and the individual mutation rates of the markers where they mismatch. The FTDNATiP™ returns an estimate of the probability their MRCA lived within the last 100, 200, 300, 400, 500, and 600 years. The FTDNATiP™ uses the most recent estimate of the mutation rate of each marker. Sample FTDNATiP™ calculations are shown in Table 1 for two hypothetical group members who have a single mismatch at **DYS385a** for the Family Tree DNA 37- marker test, and who are separated by a genetic distance = 1.

Note that because a test panel includes a range of mutation rates, it is possible for three people, all separated from each other by the same genetic distance, yet who mismatch on different markers, to obtain different MRCA estimates when any two of them are compared. For example, if two people mismatch on only one of the more slowly changing markers, the probability they will have an MRCA within a certain number of generations is much lower than that of two people who mismatch on a faster changing marker, even though both pairs would have a genetic distance = 1.

A genealogist must have done research on his family. Even someone new to the hobby usually knows his parents and grandparents. A calculation might show a high probability that two people were related within the past three generations, but if they know through their family research that they are not related within this time period, it does not make sense to include these generations in their MRCA calculation.

Fortunately, it is simple to include information you already have when you calculate an MRCA because mutation rates are independent of time. The probability that a father/son pair has one mismatch is the same as a son/grandson differing by one mismatch, and is the same as a grandson/great-grandson differing one mismatch and so on because each of these events is independent of the other. The cumulative probabilities of a mutation happening over a given number of generations is also the same, independent of which generations you include in your calculation. It does not matter if you calculate the cumulative probability of a mutation over the past 100 years starting at the present, or over the previous 100 years starting 200 years ago. The probability that a mutation occurred in that time period is the same.

Suppose you have no prior documentation regarding your relationship to someone else and you are interested in calculating what the probability is that your MRCA lived between 200 and 300 years ago based on a genetic distance = 1. Assume that as in the previous example, the FTDNATiP™ estimator returns a 27.28% probability that you two are related within the last 100 years, 57.84% probability that you are related within the last 200 years, and 77.9% probability you are related within the last 300 years. (See the example in Table 1). Starting your "clock" at the present, the chance that your MRCA lived between 200 and 300 years ago is 77.9% - 57.84% = 20.06%.

Now suppose you find documentation that proves that you do not have a common ancestor with the other person within the last 200 years. The probability that the MRCA lived within this time period is now zero, but the probability that the mutation which differentiates you took place in the last 200 years is still the same. Your search for your MRCA begins not at the present but at 200 years ago. If we were to enter this information in the calculator, then the entries for the 100 year and 200 year periods would both be 0%.

However, since there is still the probability that the mutation took place during that time period, the probability that you share a common ancestor in the next 100 year period after that (200-300 years ago) is greater than the probability originally calculated for a common ancestor in the last 100 years. If we tell the FTDNATiP™ calculator to

Table 1. MRCA calculation using the FDNATiP™ calculator for two hypothetical group members separated by a single mutation on DYS385a.

FTDNATiP™ Report

Family Tree DNA Time Predictor*

Version 1.1 - Patent Pending

In comparing 25 markers, the probability that Member1 and Member2 shared a common ancestor within the last...					
100 years is	**200 years is**	**300 years is**	**400 years is**	**500 years is**	**600 years is**
27.28%	57.84%	77.90%	89.07%	94.79%	97.58%

take this into account, we get the probabilities displayed in Table 2. The estimator returns different results from before because the probability that a mutation occurs is independent of when you start measuring it.

Table 2. Revised MRCA calculation using the FTDNATiP™ calculator for two hypothetical group members separated by a single mutation on DYS385a, assuming they do not have a common ancestor within the last 200 years.

FTDNATiP™ Report

Family Tree DNA Time Predictor*

Version 1.1 - Patent Pending

Knowing that Member1 and Member2 could not have had a common ancestor in the last 8 generations, their 25 marker comparison shows that the probability that they shared a common ancestor within the last...					
200 - 300 years is	**400 years is**	**500 years is**	**600 years is**	**700 years is**	**800 years is**
46.56%	73.24%	87.14%	93.99%	97.25%	98.76%

MRCA CALCULATIONS

Family Tree DNA FTDNATiP™ calculator gives you just what you need to convert your DNA results into information that is meaningful genealogically. Starting with your haplotype, the estimator will calculate your genetic distance from all other group members and translate your results into the probability that you have an MRCA within a specific period of time.

For more insight into the role that the number of markers, mismatches, and mutation rates play in estimating the time period of an MRCA, you may want to go through the calculations on your own and experiment with the variables to see their effects on the results. The knowledge you gain can be valuable in making a decision to test on more markers, or in interpreting your matches with new members of your group.

Research is going on to find new markers and to determine mutation rates more accurately. These will be among the many new discoveries that will influence the future of DNA testing.

An MRCA calculation depends on three variables: the number of markers tested, the number of mismatches observed, and the mutation rates. The result of an MRCA calculation is the cumulative probability that the most recent ancestor you share with someone else lived within a certain number of generations in the past. For a given number of markers and mismatches, the fewer the number of generations taken into account, the lower the probability you shared an ancestor within that time period. The greater the number of generations, the higher the probability.

The number of markers used for the MRCA calculation depends on which DNA test you and the person you are comparing with have taken. The number of mutations is the number of mismatches between you. Since the Y-STR mutation rates used by genetic testing companies are usually considered proprietary, only average mutation rates are available to genetic genealogists interested in performing their own calculations.

The basic features of an MRCA calculation are given here with a few important

X-RATED

Turner's syndrome is a genetic disorder in females in which one of the X chromosomes is missing. It is referred to as *monosomy X*, *45,X* or *X0*, since the patient only has 45 chromosomes. The X can be inherited from either parent. Turner's syndrome patients are never male since an embryo with only a Y chromosome is incapable of survival.

About 98% of all fetuses with Turner's syndrome are spontaneously aborted, with Turner's syndrome accounting for 10% of all spontaneous abortions in the U.S. It is present in 1 in 2,500 to 1 in 3,000 live births.[2] The symptoms of Turner's syndrome vary, and can include short stature, infertility, a webbed neck, skeletal abnormalities, heart defects, and kidney problems.[3]

Mosaic Turner syndrome, where some cells have two X chromosomes but others have only a single X chromosome, is also possible. In such cases the symptoms are usually less pronounced.

Continued on p. 76

X-Rated (cont)

Studying *imprinted genes* on the X-chromosome of Turner's syndromes patients has produced insight into the existence and function of various sex-related characteristics. Imprinted genes are expressed differently depending on whether they are inherited from the mother or from the father. The only organisms known to have imprinted genes, rather than whole chromosomes, are placental mammals and a plant, *Arabidopsis.*[3]

Research by Dr. David Skuse at the Institute of Child Health in London has indicated that a Turner syndrome patient who has inherited her X from her father had better social skills than a patient who has inherited them from her mother. This disparity offers clues to why boys, who inherit their single X chromosome from their mothers, are more vulnerable to disorders that affect social functioning.[4]

1. "Turner syndrome", http://n.wikipedia.org/wiki/Turners_syndrome.
2. "Turner syndrome", http://goldbamboo.com/topic-t1808.html.
3. http://www.ich.ucl.ac.uk/cmgs/seximp99.htm
4. "Social Smarts", *Discover Magazine*, **26**(10), p. 42. October 2005.

results to guide you through the discussion in the chapter *How Many Markers? How Many People?* We'll leave the gory details of the math for later. They appear in the Appendix for the die-hards.

No Frills Questions & Answers

1. *Can DNA testing tell me exactly how I am related to someone else?*

DNA testing can only tell you the probability you are related to someone else within a given number of generations. It cannot tell you the exact relationship.

2. *Why are DNA results stated in terms of probability? Is this because DNA results are not accurate?*

The interpretation of your DNA results depends on identifying mutations you have in relation to someone else. Mutations are a natural phenomenon that occurs in all types of DNA in a statistical manner. That is, we can only predict the probability that a mutation will occur within a time interval. We cannot predict the time of its occurrence. Your DNA results are accurate measurements of your DNA profile or haplotype. Probability comes along in the interpretation of the meaning of those results.

3. *What is the relationship between the number of mismatches observed between two people and the probability they have an MRCA within a certain number of generations?*

The closer two people match, the more recently they probably shared their most recent common ancestor; the more mismatches the less recently. If they have more than a few mismatches, they are not considered as being related within a time period that makes sense to genealogical research.

4. *How many markers can two people mismatch on and still be considered related?*

The number of mismatches two people can have and still consider themselves related depends on the number of markers they are tested on. The greater the number of markers, the greater the number of mismatches allowed. Two people are defined as "recently" related if they have a 50/50 chance they have an MRCA within the last 200 years, and "distantly" related if there is an even chance of an MRCA within the last 500 years. Table 3 shows how many mismatches two people can have for different numbers of markers and still be considered related using average mutation rates for each panel of markers.

5. *What if I increase the number of markers I test on but do not find any more mismatches with someone else in my group? Does this mean we are more closely related?*

Yes, probably. If you increase the number of markers you test on, but don't discover any more mismatches, you two were probably related within the more recent past than you thought when you tested on fewer markers.

6. *What if I increase the number of markers I test on and I find more mismatches with someone else in my group? Does this mean we are more distantly related?*

Table 3. Number of mismatches for haplotypes with a 50% chance of being related in the last 200 and 500 years.

N = number of markers	12	25	37
R = average mutation rate	0.0039	0.0044	0.0053
k = mismatches (200 yr.)	0.08	1.25	2.42
k = mismatches (500 yr.)	1.21	4.14	7.14

XXX-RATED

Human males have an XY chromosome pair, females have an XX pair. Each has a total of 46 chromosomes. This accounts for everyone, right?

Wrong!

Kleinfelter's syndrome is a group of chromosomal disorders in which a male has at least one extra X chromosome. The XXY chromosome arrangement is one of the most common genetic abnormalities, occurring in about 1 in 500 male births. People with this disorder are sometimes referred to as "XXY males" or "47,XXY males". Variants are much less frequent, with 48,XXYY and 48,XXXY occurring in 1 out of 50,000 male births. The most severe form of the disorder, 49,XXXXY occurs in between 85,000 to 100,000 male births. The effects on physical and mental development increase with the number of extra X's. Each extra X reduces the IQ by 10–16 points, with language most affected, particularly expressive skills.[1]

In mammals with more than one X-chromosome (normally females), most of the genes on one of the X-chromosomes are inactivated early in embryonic development, but some remain active. In males with Kleinfelter's syndrome, the remaining active genes on the inactive X-chromosome usually have corresponding genes on the Y-chromosome. These triploid genes in XXY males may be responsible for symptoms associated with the disorder.[2]

Occasionally, a variation on Kleinfelter's syndrome may occur, called an XY mosaic, where some of the cells are XXY and the rest are XY. The majority of these individuals have symptoms similar to those who are completely XXY and require the same treatment, although there are reports that mosaics have a slightly increased likelihood of fertility.[3]

1. J. Visoosak and John M. Graham, "Kleinfelter's Syndrome and Its Variants", OrphaNet Encyclopedia, http://www.orphanet.net/data/patho/GB/uk-KS.pdf.
2. http://en.wikipedia.org/wiki/Klinefelter%27s_syndrome.
3. http://www.47xxy.org/Chromosomal.htm.

It depends. Testing on more markers allows you to have more mismatches with someone else for the same degree of relationship, but only up to a limit. Table 2 tells you what that limit is. For example, if you test on 25 markers with someone and find you have 1 mismatch, you are considered closely related to that person. When you upgrade to the 37-marker test, if you show 3 mismatches you are still considered closely related. If you show 5 mismatches, you are distantly related. If you show over about 7 mismatches, you probably do not have a common ancestor within a genealogically meaningful time period.

7. *So what is the use of testing on more markers?*

If you are the only one in your group who tests on additional markers, there is little you can learn until someone else in the group does, too. There must be at least one other person you can compare your upgraded results to for your expanded test results to be meaningful.

The more information you have, the more you can know. Testing on more markers more clearly defines family relationships. As the number of people in your group testing additional marker increases, various family lines will be teased apart, giving you a more detailed family tree.

8. *Can a 12-marker test provide useful information?*

Sure, especially if you have a rare surname or a rare haplotype. A 12-marker test is most useful in ruling out a relationship. For a common name or a common haplotype, a 12-marker test might produce a large number of matches among people who are not closely related, but if someone shows a mismatch or two with the rest of the group, it rules out a relationship to a high degree of probability. In the case of a rare surname or a rare haplotype, such a rule-out test is valuable since an individual is either related to the genetically close family group or he is not.

9. *Does a nonpaternity event imply an illegitimacy in the family?*

Not necessarily. A nonpaternity event can be due to a name change or adoption.

10. *Can DNA testing always spot a nonpaternity event?*

DNA will not indicate a nonpaternity event if an illegitimacy is due to a liaison a woman had with her husband's close relative, his brother, for example. DNA will also not be able to identify the adoption of a child by his uncle or other close relative.

MRCA Lookup Tables

Since the leading DNA testing companies offer testing options based on only particular numbers of markers, and because more than a small number of mismatches is not meaningful, it is possible to use a lookup table to obtain an MRCA estimate. If you would rather experiment with the effect of different numbers of markers and mismatches or with different mutation rates, you can refer to *Appendix A* which contains more mathematical details of MRCA calculations. You can also visit the www.forensicgenealogy.info website where there are spreadsheets available you can download and use to input your own set of values.

Tables 4 and 5 are lookup tables that you can use to find the number of generations associated with a 50% chance and a 95% chance, respectively, of finding the MRCA of two individuals using specific numbers of markers and assuming a relationship within the last 500 years. The average mutation rate for each number of markers can be found in Table 3.

To use Table 4, choose the column representing the number of markers tested. Go down the column to the row representing the number of mismatches of interest. The number in the cell at this intersection tells you that there is a 50% chance of finding the MRCA as recently as this many generations ago. Repeating the same procedure with Table 5 gives you the number of generations between which there is a 95% chance of finding him.

For example, if there is one mismatch on a 25-marker test, there is a 50% probability that the MRCA will be found within 12 generations, with a 95% chance of finding him between 1 and 23.1 generations ago.

Some of the boxes in each table have been left empty where the number of generations is so great that the results are meaningless to genealogists interested in tracing their family trees. Genealogical documentation usually exists only as far back as five hundred to a thousand years.

ONLINE DATABASES

There are many reasons why you might match someone with a different surname. Every family tree has nonpaternity events, where through adoption, name change, or illegitimacy someone had a surname different from his biological father. Since surnames have been in common use for only a few hundred years, there is also the possibility that you share a genetic profile with someone without sharing their family name

Table 4. Number of generations in which there is a 50% probability of finding an MRCA versus mismatches on a 12-, 25-, and 37-marker test. Average mutation rates used are in Table 3.

		Number of Markers		
	50%	12	25	37
Number of Mismatches	0	7.4	4.95	1.76
	1	17.9	12	4.27
	2	28.55	19.1	6.82
	3		26.25	9.37
	4			11.93
	5			14.5
	6			17.06
	7			19.63
	8			22.2

Table 5. Generations between which there is a 95% probability of finding an MRCA for various numbers of mismatches on a 12-, 25-, and 37- marker test. Lower end of range represents 2.5% probability and higher end of range represents 97.5% probability. Average mutation rates can be found in Table 3.

		Number of Markers		
	95%	**12**	**25**	**37**
Number of Mismatches	**0**	0.3 - 39.2	0.1 - 15.3	0.06 - 9.4
	1	2.6 - 59.3	1.0 - 23.1	0.6 - 14.2
	2		2.6 - 30	1.6 - 18.4
	3			2.8 - 22.3
	4			4.3 - 26.1
	5			5.9 - 29.7

because you descend from close relatives who adopted different surnames. Think of them as genetic cousins.

As DNA testing is becoming more popular the chance of two people matching who have different surnames is increasing rapidly. It is becoming ever more important to widen your search beyond your own study group.

The increase in the popularity of DNA testing has been accompanied by the establishment of online databases that allow you to compare your haplotype to thousands of other haplotypes. Along with the name and contact information for any matches that are found, these sites also give you information on haplotype and marker value frequencies and the geographical distribution of participants. Some will predict haplogroup membership based on statistical analysis of the Y-STR composition of various haplogroups. Although many of the websites that contain these databases are sponsored by major DNA testing companies, they are accessible to everyone.

Websites with the most popular databases are listed below.

www.ysearch.org provided by Family Tree DNA accepts Y-STR results from all the other testing laboratories for comparison with its database. It also accepts family trees in GEDCOM format. As of September 2005, the Y-Search database contained 14,778 surnames, 12,395 unique haplotypes and 16,257 records. To access the database, a user must create a free account by submitting his Y-STR results and providing information on his most distant known ancestor along the direct male line. The number of markers and mismatches to search are user-selectable. A search can be limited on the basis of surname, surname variants, geographical location, and haplogroup. A user can also search on a specific surname or user ID. There are Y-search Compare and Genetic Distance™ Report facilities for comparing specific user IDs in the database.

www.smgf.org, sponsored by the Sorenson Molecular Genealogy Foundation, correlates genetic and genealogical information. As of September 2005, it contains 9,446 surnames associated with 11,095 unique haplotypes. The www.smgf.org site provides a drop-down menu where you can enter your haplotype for easy comparison with the database. Closely matching haplotypes are returned with a table indicating the status of the match for each marker, but without giving the actual value of any of the markers. The table is linked to the pedigrees of each matching haplotype and to an estimate of the MRCA for each match. Names of living individuals are not provided with the pedigrees for privacy reasons.

To add your haplotype to the SMGF database you must request a free GentiRinse test kit and return the samples to the Sorenson laboratory. No personal results are returned. All results are added to the publicly accessible database on the site, but personal information on haplotype composition and vital statistics (name, birth date, etc.) are not provided with the match information. Sorenson actively recruits individuals with extensive genealogies, family surname groups, and members of specific populations. A list of surnames and geographical locations that are in the database is given,

IQ XY

Firm evidence shows that although men and women have physiological differences that translate into variations in cognitive abilities, the genders are well matched on average intelligence.

Yet differences emerge when the distributions of IQs is examined instead of the averages. While females average slightly better on IQ tests, more men are mentally retarded, and more men have IQs at 135 and above.[1]

The logical place to search for a genetic cause of gender-related intelligence differences is the X chromosome. The Y-chromosome does not offer good hunting ground, since the few genes it carries are related to sperm production and physical characteristics associated with being male. Only 54 of the 1,098 protein-coding genes on the X seem to have functional counterparts on the Y. When a gene on one X chromosome mutates in a woman, a backup gene on the second X chromosome can fill the gap. But when a gene on the X chromosome mutates in a man, his Y stands idly by, like an onlooker at a train wreck. X-related genetic mutations have

Continued on p. 85

along with the statistics, the physical properties of, and references to scientific research papers for each marker.

www.y-base.org is sponsored by DNA Heritage. It provides a user-friendly way of comparing haplotypes to their database and for submitting your haplotype through a drop-down menu of marker values. There is a search engine based on both haplotype and surname that provides the contact information of contributors. The site also provides a "haplomatic" for predicting your haplogroup from your STR profile, along with statistics on the allele values and haplogroup composition of the database.

www.yhrd.org - The Y Chromosome Haplotype Reference Database was created by the Forensic Y-User Group. This database is a successor to the www.ystr.org database. The website provides haplotype frequency estimates for both forensic and genealogical use, and also for wider studies of population composition. As of September 2005, the database contains 32,192 haplotypes drawn from 271 populations worldwide.

The www.yhrd.org website provides useful information on haplotype frequencies, the statistical distributions of the various markers' allele values, the mutation rate of the Y-User Group-defined core and extended Y-STR markers, and the position of each of these markers on the Y-chromosome. It is possible to search the database for anonymous matches to a haplotype and to obtain information on the geographical location of its donor. The site only accepts submissions from laboratories that pass a quality control exercise involving the evaluation of blind samples supplied by the Institut fur Rechtsmedizin, Humboldt-Univeristat, Berlin, Germany. There is a list of contributing laboratories on the site with contact information.

www.cstl.nist.gov/biotech/strbase/index.htm was created by John M. Butler and Dennis L. Reeder of the National Institute of Standards and Technology (NIST) as a repository of information relating to the use of STRs in human identity testing. The site covers many more markers than those in the core and extended STR sets. It has an excellent library of information on STRs more suited to the experienced researcher, including information on the properties of the markers, and population data.

www.mitosearch.org is provided by Family Tree DNA for matching mtDNA results. Use of the database requires the creation of a free account through the submission of mtDNA results. Searches can be performed on the basis of mtDNA SNP results or on haplogroup. The site also accepts family trees in GEDCOM format. Family Tree DNA recommends that customers

IQ XY (cont)

been linked to an estimated 300 genetic diseases and disorders in males, including color blindness, muscular dystrophy, and about 200 brain disorders.

Another reason to link intelligence with features on the X-chromosome, according to Gillian Turner, professor of medical genetics at the University of Newcastle in Australia, is that the X-chromosome provides a means of quickly distributing a gene through a population, and "no human trait has evolved faster through history than intelligence."

If an important gene were carried on the Y-chromosome, only males would inherit it. But for the same gene on the X-chromosome, everyone has a 50/50 chance of inheriting it from his mother, and a female would have a 100% chance of inheriting it from her father.

The good or bad fortune of the X-chromosome is good or bad news for males. This is especially true for genetic disorders affecting the brain. According to physician and human geneticist Horst Hameister and his group at the University of Ulm in Germany, 21% of

Continued on p. 86

IQ XY (cont)

brain disabilities map to X-linked mutations.

Hameister believes that just as males are more prone to genetic disorders caused by X-related mutations, they benefit more from genetic advantages. While females have two Xs, males get their X exclusively from their mother. When a male inherits a favorable X-related genetic mutation, there is negligible chance it will be overridden by a corresponding gene on his Y-chromosome. In a female, a good mutation on one X might be overridden by a normal gene on her other X.

This has important consequences for the next generation. A man who inherits a favorable X-mutation from his mother can not pass it to a son, although he will always pass it to a daughter. Since the daughter's paternal X-chromosome recombines with her maternal X-chromosome in the formation of the daughter's egg cells, there is a 50% chance that she will pass it to the next generation.[1]

1. This summary contains quotes from the fascinating article "X-Rated" by Ellen Ruppel Shell that appeared in *Discover Magazine, Special 25th Anniversary Edition*, **26** (10) 42-43, October 2005.

testing the full mtDNA sequence compare their results to other records using the BLAST search engine on the NCBI (National Center for Biotechnology Information) website. This search engine is linked to your personal page on the Family Tree DNA website. To access the BLAST search, go to http://www.ncbi.nlm.nih.gov/BLAST/ and click on the link "Quickly search for highly similar sequences (megablast)."

REFERENCES–ARE WE REALLY COUSINS

1. S. Siguraordottir *et al*, "The Mutation Rate in the Human mtDNA Control Region," *Am. J. Hum. Gen.* **66**:1599 - 1609 (2000).

WOW, OW, WO, AND O STORIES

Every genealogist hears those thunderclaps known as WOW moments when his hard work pays off in a big way. After a long drought, a WOW experience can renew the interest of even the most jaded among us. One of the real values of DNA analysis is that it opens up new sources of WOW in unexpected directions.

There are several kinds of WOW experiences. A WOW moment can happen when you stumble across something that ties up a lot of loose ends. It can also put your research into a new perspective by offering an unexpected connection in the present that you have no explanation for in the past. It might also give you hints of a relationship in the past that you cannot yet find proof for in the present. A WOW moment occasionally resolves a mystery, but more than likely it will present new questions, sending you along a new path on your search for enlightenment.

A WOW experience can bring resolution to a long suspected truth. The example of Mr. Bell who resolved the family story of his relationship to the Mumma family through a DNA match is a great example of a WOW experience. (See the chapter *The Ins and Outs of Genetic Genealogy*). The beginning of Mr. Bell's story was that his great grandfather, said to have had the surname Mumma, died before his grandfather was born. His grandfather was then adopted by his great grandmother's second husband named Bell. The end of the story was that DNA analysis along with written documentation showed that his great grandfather was really a traveling salesman named Mumma who was later married to someone else with whom he had several children. That was how Mr. Bell came to be known as Mr. Bell and not as Mr. Mumma.

If a WOW story shows an unexpected relationship in the present that cannot yet be explained in the past, it is called an OW story. OW stories are very common in genetic genealogy and occur when two people unexpectedly match closely, without a known family connection. For example, in 2002 a Catholic priest from New Jersey joined our Fitzpatrick study only to discover his DNA results were wildly different from the rest of the group's results. Seventy out of seventy-three of us form a homogeneous group, matching about 20 out of 25 markers, belonging to the R1b haplogroup. But Fr. Fitzpatrick's DNA results show that he matches on the average 7 markers with the rest of the group, and indicate that he has a haplotype associated with the I1c haplogroup.[1]

In 2004, this situation took a surprising turn when a Fitzpatrick from New South Wales, Australia matched Fr. Fitz exactly. The two men were unaware of each other. By comparing genealogies, we found that their families came from two towns that are ten miles apart on the west coast of Ireland. The priest's family had come to the U.S. as famine immigrants in the 1850s; the Australian's had arrived in Australia in about 1900. How could they have the Fitzpatrick surname and belong to such a different haplogroup from the rest of the group members?

We have developed some theories as to how this came about. The I1 haplogroup is thought to have spread into Europe from the Iberian Peninsula after the last glacial maximum. The I1c subgroup had France as its source[2] and spread first into the Germanic and Scandinavian countries and then into the

British Isles. One possible reason for the membership of these two Fitzpatricks in such a distinct haplogroup is that they descend from a Viking who left a genetic souvenir during a pit stop on his way through coastal Ireland, but we will probably never know for sure. Until the mystery is solved, these two Fitzpatricks are the substance of an OW story.

If a WOW story indicates a relationship in the past but no relationship in the present has been established, it is called a WO story. WO stories come about, for example, when two people with the same surname have the same genetic disorder but have not yet determined their relationship.

Several years ago, my contact with a woman named Maggie led to a rare combination WOW–WO experience. She wrote me to investigate a possible relationship with my maternal grandmother whose maiden name was Bernard. Among the few facts she knew about her great grandmother Marie Bernard was that she had been married to a naval officer, and had traveled with him to Japan in the late 1890s. Maggie knew that her grandfather Charles was born in Colorado, but had no explanation why Marie would have visited there. She knew Marie had been born either in France or in New Orleans, and she had heard that there was a painter somewhere in Marie's family.

Since the surname Bernard is very common among the descendents of French immigrants in New Orleans, I realized it would be very hard to establish a relationship between Maggie's family and my grandmother's. But I promised I'd look through my notes. I usually copy down info on all the names in my family that I come across, and not just the names of people I know I am relate to.

To my surprise, I discovered a note I had written to myself twenty years ago about a Bernard family I had noticed in the 1870 New Orleans census: Francois Bernard, portrait painter, Victorine, his wife, Marie his 13-year-old daughter, and Charles, his 9-year-old son. Surely this was Maggie's family, WOW!

Over the next few weeks, Maggie and I exchanged excited emails about her famous great great grandfather. Francois Bernard was a portrait painter who was brought from France by wealthy plantation owners in Southern Louisiana to paint their portraits. He was a contemporary of Edouard Degas, the well known French painter who, for a while, lived in New Orleans with relatives. Some of Francois' portraits now hang in the Louisiana State Museum in Baton Rouge.

This was a lot of fun, but our communications became serious one day when Maggie told me the reason she had become involved in genealogy was that she had a genetic disorder called Charcot Marie Tooth (CMT) syndrome. As she investigated where in her family it could have originated, she was able to rule out all of her family lines except for the Bernards.

CMT is a neurological disorder named after the three physicians who first characterized it in 1886, Jean-Martin Charcot and Pierre Marie of France, and Howard Henry Tooth of the U.K.[3] CMT is manifested by a slow deterioration of the nerves in the extremities. Once the nerves have been damaged, the muscles begin to atrophy. The symptoms of mild cases are extremely cold hands and feet, and tingling in the extremities. Moderate cases are characterized by carpal tunnel syndrome, shin splints, hammertoe, and rheumatoid arthritis. In more severe cases, the person may develop drop foot and might even be confined to a wheelchair. It can be inherited by both men and women in three ways: 1) it can be passed by a mother to a son or a daughter via the X-chromosome, 2) it can be inherited as an autosomal dominant gene, so that a person needs only one copy of the defective gene to be afflicted, or 3) it can be inherited as an autosomal recessive gene, so that a person must inherit a defective copy of the gene from both parents in order to manifest the disease.

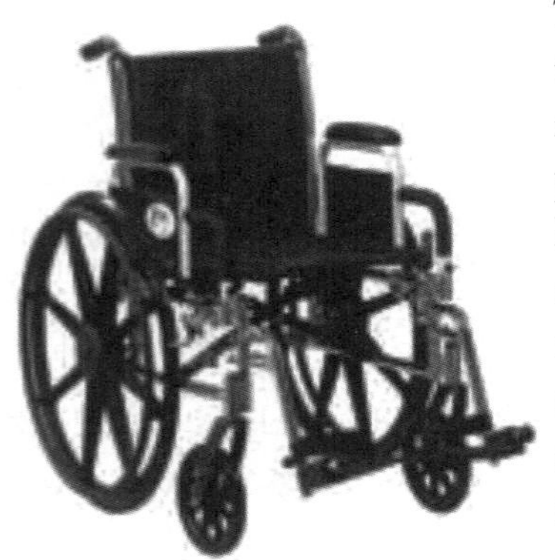

Maggie told me that she had a moderate case. She could still drive, but the next day she had to take it easy because his hands hurt from clutching the steering wheel. She walked with a cane and limited herself to only a few short emails a day because of the pain she experienced in her hands from typing.

As Maggie recounted her symptoms to me, I realized that I had the identical symptoms in my family–but only along the Bernard line. My grandmother had hammertoe and severe rheumatism; my mother and I are both noted for having extremely cold hands and feet. Other symptoms are prevalent along the Bernard line of our family, including severe carpal tunnel syndrome.

Upon further investigation, I discovered that there is a high prevalence of CMT among the descendents of French immigrants who live Southern Louisiana. This group includes many Bernards.

Based on only the fact that we have the same genetic disorder, Maggie and I might not have suspected that we had a common ancestor. But because CMT appears along lines of our families that share the Bernard surname, because we come from the same ethnic background, and our ancestors were living in the same geographical area at about the same time, there is a reasonable probability that we are related through a distant Bernard who first hosted the CMT mutation. This is a WO story, with strong suggestions of a connection in the past, but no documented connection in the present. Hopefully through DNA testing in conjunction with paper documentation, we will be able to establish a family link so that our WOW–WO story will become WOW–WOW.

There are discoveries of a unique genetic feature in a family that are worth investigating further both in the past and in the present. These are the substance of O stories. Unlike my experience with Maggie, an O experience provides little information suggesting potential relatives outside of known family members, and minimum evidence of how you might connect to others who are found with the same unique feature. But because the feature is genetic, it must have been handed down in your family. If you find someone that has the same characteristic, if it is rare enough, there is a high probability that you are related, even if DNA testing is not possible because you are not connected through the direct male or direct female line.

When Mr. C. underwent medical tests to better diagnose the cause of his heart attack in 2004, he was presented with the bad news that he had cancer somewhere in his body.[1] Several months later, he underwent colo-rectal surgery, followed by chemotherapy and radiation. In 2005 his condition worsened when he experienced internal bleeding. He spent almost all of July and August in and out of hospitals and emergency rooms, and was finally sent in mid-July to a research hospital.

When the medical experts at the hospital could find no reason for his bleeding, Mr. C. was given a small encapsulated TV camera to swallow that allowed a view his intestines. There was no signs of cancer, but the TV images showed pox-like marks in the lining of his small intestine. The condition had no name and no known cure. It could be treated with surgery to some extent and by giving transfusions to replace the blood he was losing. The research team began investigations that turned up one person in India and about 30 people in Italy who were just discovered to have a similar problem.

Mr. C's family medical history and family folklore hint that the condition may be genetic. The Z line of Mr. C's family appears to have originated in the Rheinland. His 5th great grandfather Johannes Z came to the U.S in 1753, first settling in eastern Pennsylvania, his descendents moving farther west to Ohio and Iowa. Johannes' son Adam married Christina B whose family apparently originated in Basel, Switzerland and Strasbourg, France. The family was Lutheran, but family folklore tied them to a Jewish ancestry.

The earliest existing photograph in Mr. C's family is that of his great great grandfather, Jacob Z, a severe, dark man with a long curly beard. Mr. C is also very dark and is constantly being asked if he is of Italian or Middle-East ancestry. According to information that has been handed down in the family, Jacob was supposedly given the wrong medicine and died. Some family members passed on the story that he may have bled to death. Other members of this family line who have suffered from chronic anemia are Mr. C's grandmother Edna (granddaughter of Johannes), Edna's daughter Amy (Mr. C's mother), Mr. C's sister and his brother's daughter.

There are many possible reasons for the appearance of the genetic disorder in both the males and the females in the Z line of the C family, especially since the occurrence of the disease is rare. The defective gene could be a gene handed down on the X-chromosome, so that it is possible for both the males and the females in the family to inherit a copy. It is also possible that since the Z family lived in the same area for several generations, there could have been sufficient intermarriage to pass the gene to a high percentage of the population. If a man and his father both suffer from the disease, it could be coincidental. It could be that the man's mother and his paternal grandmother were both carriers, causing it to appear that the gene resided on the Y-chromosome when it was actually present on the X-chromosome.

Will the family ever find the cause of the C family's anemia? Are the stories about Johannes' bleeding to death true? How do Mr. C's recent problems relate to those of the recently discovered Italians? Does he share a common ancestor with them? What about the person in India who has the same medical problem?

Until further information is discovered about the C family's medical condition and genealogical research is done to investigate a possible connection with the Italians who suffer from the same disorder, these questions will remain unanswered, and the story will remain "O"-pen ended.

REFERENCES–WOW, WO, OW, AND O STORIES

1. http://freepages.genealogy.rootsweb.com/~gallgaedhil/haplo_i1c_part_2.htm
2. Siiri Rootsi et al, "Phylogeography of Y-Chromosome Haplogroup I Reveals Distinct Domains of Prehistoric Gene Flow in Europe," *Am. J. Hum. Gen.***75**:000-000 (2004).
3. Facts about Charcot-Marie-Tooth Disease and Dejerine-Sotta Disease, http://www.mdausa.org/publications/fa-cmt.html#whatis

NONPATERNITY EVENTS

One of the eventualities of every surname DNA study is the appearance of someone whose haplotype is completely different from all the others in the group. This is referred to as a non-paternity event, a situation where through adoption, a name change, an illegitimacy, or any combination of these, an ancestor did not have the genetic profile of his alleged father. Sometimes the reason for the discrepancy is buried long in the past, going unnoticed for generations. Sometimes it is more recent and can be tracked down through testing others who, at least on paper, belong to the same family line. The famous study of the descendents of Thomas Jefferson and his house slave Sally Hemmings is a good example of a situation where verifying a non-paternity event was the focus of a study and not an exception to expected results.

There are many complications in figuring out the cause of a non-paternity event. Before the mid-1900s adoptions were often informal; that is, if a child was orphaned he would go live with the next door neighbors or with his maternal uncle Fred and nothing more would be said about it. Often adoption was considered a "family secret," so that even a child who was adopted legally would not be told about it.

mtDNA INHERITED FROM THE FATHER

Since mtDNA carried into the egg by a sperm normally disappears shortly after conception, it has been assumed that a child inherits mtDNA exclusively from his mother, so that mtDNA is handed down along the exclusively female line of a family. It has been assumed that while a man inherits his mtDNA from his mother, he does not pass it on to his children.

In one case, this assumption was shown to be incorrect. In 2002, researchers at the University Hospital Rigshospitalet in Copenhagen reported a case involving a 28-year-old-man suffering from a disorder impairing the ability of his muscles to metabolize adequate amounts of oxygen. The disorder caused him to become very fatigued when he exercised, although his heart, lungs, and muscles appeared normal.[1]

His muscle mtDNA was compared with mtDNA from the blood of his parents, his paternal uncle and with the blood and muscle tissue of

Continued on p. 97

Immigration is one of the most frequent occasions where a name change can occur. Before the creation of the Ellis Island immigration station in 1906, there were many reasons why an immigrant was admitted into the U.S. under a name different from the one he had in his old country. Surname changes were common, especially among immigrants when a change in language was involved. Often, an immigration official did not understand the accent of an immigrant and recorded what he thought he heard (Burdo for Bordeaux), or the immigrant gave a translation of his name (Ash for Lafreniere, Bird for Loiseau).[1] Occasionally, an immigrant would be escaping legal problems or religious persecution and gave an assumed name.

After Ellis Island was opened in 1906, immigration name changes became far less common. According to Marian L. Smith, an historian for the Immigration and Nationalization Service (INS) in her article "American Names/Declaring Independence,"[2] passenger lists were created at the point of departure, and immigration officials worked from the ships manifests while processing arrivals. While these manifests could contain misspellings, officials were also under strict regulations that forbade them to change a name unless there was an obvious error or unless requested to do so by the immigrant himself.

Ms. Smith's article adds that, contrary to popular folklore, family name changes did not occur because of any language barrier at Ellis Island. "One third of all immigrant inspectors at Ellis Island early this century were themselves foreign-born, and all immigrant inspectors spoke at least three languages. They were assigned to inspect immigrant groups based on the languages they spoke. If the inspector could not communicate, Ellis Island employed an army of interpreters full time, and would call in temporary interpreters under contract to translate for immigrants speaking the most obscure tongues."[3]

There are other situations where a surname could change, for example, a family could have changed its name when it moved to another part of the country. Mark Haacke, the coordinator of the Haacke family study, found a branch of his family changed their name to Hickey when they moved to Oklahoma. What was originally a rare German surname was transformed into a common Irish surname, considerably complicating his family research.[4]

The rate of illegitimacy has been studied extensively and found to vary considerably with nationality, culture, and time period. By "rate of illegitimacy" is meant the number of illegitimate births divided by the total number of women capable of having chil-

mtDNA (cont)

his sister.[2] Although it was expected that the patient's mtDNA would match only that of his mother and sister, it was found that it matched the mtDNA of his father and uncle. Further investigation showed that 90% of the mtDNA in his muscle tissue came from his father, but the mtDNA in his blood, hair roots, and fibroblasts came from his mother.[3] Researchers also found that 0.7% of the mtDNA found in his muscle tissue, or 7% of the 10% of his muscle mtDNA that he inherited from his mother, contained segments of his father's mtDNA. This indicated that in very rare cases mtDNA can be recombinant, that is, it can contain contributions from both the male and female lines.

To-date this is the only reported case of the transmission of paternal mtDNA to an offspring.

1. Schwartz, M.A., et al, "Paternal Inheritance of Mitochondrial DNA", *N. Eng. J. of Med.* (**347**), p 576.
2. http://www.the-scientist.com/news/20020822/03.
3. http://www.newscientist.com/article.ns?id=dn2716.

dren. The illegitimacy rate is different from the illegitimacy ratio, which is the number of illegitimate births divided by the total number of births over the same period of time. For a small number of births occurring in a large population of fertile women, the illegitimacy rate can be low while at the same time the illegitimacy ratio can be very high.

In Catholic and Islamic societies illegitimacy rates are extremely low, but in Great Britain and Western Europe one in three children is born out of wedlock. Illegitimacy rates have been found to vary in time for Great Britain and Western Europe from 4.4% in 1540 down to about 1% in the 1600s, and in the 20th century alone, from 4% at the beginning of the century to 30% near the end. Illegitimate births are found to be higher among the uneducated and in situations where marriage occurs at an early age.[5]

Non-paternity events are especially noticeable in family studies that involve a rare surname or a group descending from only a few known ancestors. These studies involve only one or at most several related haplotypes, so that wide variations are obvious. Non-paternity events can be harder to spot for studies that have multiple sources for a surname; for example, a surname that was related to a geographical location (Hill, Rivers, Lake), a profession (Miller, Smith, Wright), or that was a patronymic (Johnson, Williamson, Jackson). In the case of the surname of a wealthy family that existed when surnames were becoming common, a nearby group could have adopted the name for economic reasons or for protection. These studies might include so many different haplotypes that the occurrence of a non-paternity event is lost in the noise.

Nonpaternity events can also be hard to spot for families that have lived in the same geographical area for many generations. In this case, a high percentage of inhabitants of the area might descend from a single common ancestor so far in the past that familial relationships have long been forgotten. If a woman has an illegitimate child, chances are significant that the genetic father of that child is a relative of her husband. If their relationship is close, for example, if the two men are brothers or close cousins, her son will have a Y-DNA profile probably identical to both men and the nonpaternity event will be genetically undetectable. If the relationship is distant, the Y-DNA profile of the child will match the biological father, but differ somewhat from that of the woman's husband. Although the child will still have a common ancestor with his "brother" (son of the woman and her husband), the ancestor will not be the husband as assumed, but someone from the more distant past. While DNA analysis of the Y-chromosome may give a suspicion of a nonpaternity event of this type, there is no certainty because of the statistical nature of the mutation process.

Nevertheless, even in a group with a complicated variety of haplotypes, some results can turn up that are considerably off-the-wall by even the most liberal standards. Since such results indicate that someone of considerably different genetic makeup was involved with a family, they can provide a interesting opportunity for investigating a family's past historical or political connections.

REFERENCES - NONPATERNITY EVENTS

1. http://genealogy.about.com/gi/dynamic/offsite.htm?site=http%3A%2F%2Fwww.rootsweb.com%2F%7Ecanqc%2Fangloabc.htm.
2. Marion L. Smith, "American Names/Declaring Independence," http://genealogy.about.com/gi/dynamic/offsite.htm?site=http%3A%2F%2Fwww.immigration.gov%2Fgraphics%2Faboutus%2Fhistory%2Farticles%2Fnames.htm
3. Marian L. Smith, op cit.
4. Mark Haacke, private communication.
5. http://web.staffs.ac.uk/schools/humanities_and_soc_sciences/census/illegit.htm.

STARTING AND MANAGING MY OWN STUDY

If you do not find a study already underway for your surname, you should consider starting and managing your own. This is not difficult thanks to the tools provided for Group Administrators.

Organizing Your Study

The first step in forming a group study is to find a Group Administrator who will oversee the project. Extensive knowledge of DNA testing is not necessary. The job of the Group Administrator is to organize the members as they join the project, go through the testing process, and receive their results.

Let's assume that you have decided to take on the responsibility of the Group Administrator yourself.

It is important to formulate your study objectives. You may decide to focus your efforts on the descendents of a particular ancestor, or to concentrate on a specific geographical region. You may be interested in certain spelling variations of your surname. Stay flexible and to be willing to modify your objectives if new information is uncovered about your family. You might want to expand your study to include other spelling variations, or start a separate study on a subgroup of participants.

Recruiting Participants

There are many places you can look for guinea pigs–I mean participants–for your study. Recruiting your immediate family members is a start, but it's best to include a diverse collection of people with your name. Family reunions, genealogy clubs, historical societies, direct snail mail and email are all good sources of participants. The Genealogy-DNA-L@rootsweb.com[1] list is a good mailing list for an initial announcement of your study. You might want to post a message to the Family Tree DNA Forum,[2] the World Family Network Forum,[3] International Society of Genetic Genealogy (ISOGG) mailing lists and forums, or GenForum.[4] Chances are you already have experience tracing your family and have been in contact with other researchers by the same surname. These people are the most likely to join your study in its early stages as they share your interest in establishing kinship.

If you decide to use online mailing lists or forums to solicit members for your group, there are a few do's and don'ts that are useful to know. Most of them are common sense items. Make sure your letter to the list is well written and follows the list's or forum's guidelines. Your message will be better-received if it is presented as a query or an outreach instead of a commercial for your group. One of the best ways to announce your study is by simply stating that you are starting a DNA study (and give the surname) with your contact information for anyone who wants to know more. If you plan to focus the study on a particular geographic area or on descendents of a specific person or group, include that in your posting. There is no need to give the gory details about costs or privacy considerations. In my experience, it is a good idea to mention that the test does not involve giving a blood sample, it only involves swiping the inside of your mouth with a cotton swab. I have found this to be one of the important misconceptions that prevents people from getting involved, best dealt with right away.

Make sure that the list or forum you are sending an announcement to is appropriate to your study's objectives–that it pertains to the surname you are studying and any geographical areas or specific ancestors you are focusing on. Also make sure that the list topic is compatible with an announcement about a DNA study. Do not copy (or blind copy) many lists with the same message attempting to blanket the community. Your posting might be interpreted as spam and you could get locked out. Update the list when there have been exciting developments or you have reached a milestone in the number of people in the study, but it is probably not necessary to send an email every week. For more great tips on how to use mailing lists to get the word out about your DNA study, see Lauren Boyd's message to the Genealogy-DNA-L@rootsweb.com list "How to Win Friends and Influence List Administrators" posted May 5, 2003.[5]

If you are a male with the surname you are studying, be one of the first to enroll in your study. If you are a male but are interested in studying your mother's surname, have a male cousin with that name take the test. If you are a female, you must coerce a brother, uncle or other male relative to participate. When I first learned of DNA studies, I was returning from our Fitzpatrick Clan2000 Gathering in Ireland. I had a ready-made source of potential participants. Shortly thereafter, my brother became the first member of our Fitzpatrick group.

For maximum benefit, select relatives who are as far apart in the family tree as you can find. They should be separated by as many *transmission events* as possible. A transmission event is defined as the passing of Y-DNA from one generation to the next, with the number of transmission events counted from the first member up to the common ancestor and then back down to the second member. This will ensure the highest probability that they will show a mutations relative to each other, allowing you to distinguish between family lines.

Since the most recent common ancestor of two brothers is their father, they are separated by two transmission events (one from the first brother up to the father, and the second from the father down to the second brother) so that their results will likely be identical. The same argument holds true for other types of close relatives such as a father and his son. Testing first cousins is a slight improvement, since their most recent common ancestor is their grandfather, separating them by four transmission events. Testing second, third, fourth cousins, and so on, increases the number of transmission events separating them and therefore the probability that they will show a mutation.

The farther apart two members of your family appear in your family tree, the more beneficial it is to include them in your study. Having two brothers take the test does not give you much information on the structure of your surname, although it may contribute to the statistical knowledge of the mutation rates for your group. You might want to include particular members of a family to help isolate where a mutation has occurred in your family.

Potential study members may have reservations about providing a DNA sample for analysis. Part of this may involve a misunderstanding of the nature of the junk DNA used for the study. It cannot be used for medical or insurance purposes. Another concern may be over privacy issues. Who is going to see the results? How can privacy be safeguarded?

These issues can usually be addressed through education about the DNA testing process. All DNA testing companies keep results strictly confidential and follow security procedures for all the tests they conduct. As sample submitted for analysis is immediately assigned an identification number that is used throughout the testing process. The identity of its owner is not known except to a very few people. Results are not shared even within the group unless permission is granted to do so. Matches within a study or with those participating in other studies are handled either by the company doing the testing or by the study coordinator, who normally has access to his list of participants and their test results. Testing companies do not share results with other organizations–the industry is highly competitive–and any release of confidential information could destroy a company's position within the community. For more information on privacy policies and procedures, visit http://www.duerinck.com/privacy.html.

WANTED DEAD OR ALIVE

In 1995, the remains of the Nazi physician Josef Mengele, "the Angel of Death," were identified through analysis of DNA extracted from remains buried in Brazil, where Mengele had fled after being captured by the Allies at the end of WWII. He was supposedly killed in a boating accident in 1979, but authorities were not sure that the man who drowned was Mengele. A comparison of the DNA extracted from the bones with that of Mengele's wife and son who were still living in Germany proved beyond a reasonable doubt that Mengele's whereabouts had finally been established.

1. http://www.wellcome.ac.uk/en/genome/genesandbody/hg07f007.html

Reluctance to joining a study may also arise from the cost of testing. There are several ways to address this issue. Some companies allow group administrators to establish general funds or scholarship programs for their projects. As new members enroll in the study, they can be approached to donate a few extra dollars to cover testing for members who could not otherwise participate. As the study grows, members will sponsor newcomers who are strategic to their own research. Point out to a prospective participant that only one member of a family needs to be tested so that he might suggest sharing the cost with other interested members of his family.

Even if someone inquires about your study, but does not join, keep in contact with him. As our Fitzpatrick study has grown, we have collected so many family trees from members and non-members that even if someone does not want to participate in our group, we have the potential to connect him into a branch of the family through a paper trail.

Remember that your surname study is an ongoing effort. There might not be a rush to join the moment it is announced; it might take time to attract members. But the genealogy community is tight-knit, especially among people studying the same surname. It helps to get the word out on breakthroughs or surprising developments, and to update the family from time to time on the study's status in terms of how many people are included and how the results are shaping up. I have had on the average just over one person a month join my Fitzpatrick study for the last five years. Sometimes, someone has joined my study years after he first heard about it. The important thing is to keep the word about your study on the street and to have patience. "If you build it, they will come."

Managing Your Project

As a Group Administrator, you will have the responsibility for managing your study members, tracking their samples through the system, and for organizing and displaying your group results. This is not difficult, and there is a lot of help available when you need it. Keep in mind that there are now about 2,500 name studies being conducted in the world, all with Group Administrators. So when you take responsibility for your own study, don't think of yourself as the MayTag repairman of genetic genealogy. You become a member of a community of Group Administrators, everyone of whom needed help at one time or another. If you get stuck, there are many people around who can help you from their personal experience. The International Society of Genetic Genealogy (www.isogg.org) is a great place to start when you have questions about running your group.

As an example of the kind of facilities available to project leaders, Family Tree DNA provides each Group Administrator with his own Group Administrator Page (GAP) on the internet. The GAP offers you all the guidance you may need to keep your project running smoothly. The areas where the GAP offers help are listed in Table 9. The Family Tree DNA GAP also offers a Quick Reference guide with brief explanations of each option and a GAP Interpretation help file on some aspects of interpreting results that may arise in discussions with your group members. These include General Concepts, Relatedness, and Special Marker Technical Questions.

The **Project Profile Page** allows you to control the content of your Project Join Page and your Project Website. These are the important first looks that a prospective member has when considering your study, so you should keep them up-to-date on project objectives and recent developments. If there have been new developments in your group, such as the discovery of a spelling variation of your name or an unexpected connection among group members, you should update your Join Page and website to maximize your chances of attracting new members. The Project Profile Page also allows you to keep up-to-date on project membership and group results by providing options for email notification of new orders and of results that have come back from the lab.

The **Member Page** contains the links you need to manage your group members. You can log in as any group member and access his contact information. It states whether the member is interested in all matches, or just matches to his surname. It displays the date his kit was received by Family Tree DNA and whether the member agreed to release his information to potential matches in the Family Tree DNA database. This page provides a link to the member's Genetic Distance Report (leading to the Family Tree DNA Time Predictor or FTDNATip™ Report), where he will be compared to all members of the project.

The **Add a member** link allows the Group Administrator to order a kit for a new participant. This requires providing contact and billing information. The kit may be ordered for one person but paid for by someone else or through tapping the project's General Fund. Each order is good for one kit. The new member's name will automatically be added to the group.

If more than one kit is required, they must be ordered by the Group Administrator through the **Order multiple kits** page. The Group Administrator can then distribute them.

Once a kit is received by Family Tree DNA, it is included in the next batch shipped to the University of Arizona laboratory for analysis. Batches are shipped every week. A Group Administrator receives an email when his group's kits are sent out to the lab. The link to **Pending shipment to lab** indicates the date a kit is received and that it is waiting to be sent out.

Unreceived lab results and **Received lab results** give all the information required to track a kit through the system once it has left Family Tree DNA's administrative office. These two pages indicate the kit number, product, which lab test has been ordered, and either the estimated completion date and batch number for a kit that is in

Table 9. Family Tree DNA Group Administrator Page (GAP)

Project Profile Page/Send join authorization	Reviews and allows edits to project page, manage email notifications options, and send a link to the Sign-up page for the group
Member information and genetic distance reports	Lists members with basic information about each: Kit No., Code No., Status of kit and release form, link to genetic distance report, and more.
Add a new member by placing an order for them	Leads to the Sign-up page for the group.
Order multiple kits for redistribution	Leads to GA's multiple-kit order page
Pending shipment to lab	Lists kit no., product, lab test, name and comments for each kit that has arrived at FTDNA and is pending shipment to the lab
Unreceived lab results	Lists kit no., product, lab test, name, batch no., estimated result date, and comments for each kit that has not yet finished lab analysis
Received lab results	Lists kit no., product, lab test, completion date, and name for each set of results that has been reported
Generate Y-DNA Results for copy & paste	Y-STR results in table format suitable for copy and paste to another web site, to an Excel spreadsheet or to a Word document. Includes links to download to Excel or to convert to pdf format. Includes haplogroup prediction or SNP test results
Generate mtDNA Results for copy & paste	mtDNA results in table format suitable for copy and paste to an Excel spreadsheet or to a Word document
Unique Haplotypes	Unique haplotypes for the 12-Marker, 25-Marker, and 37-Marker test in table format
Family Project Website Setup	Sets up and allows input to the Family Project Website, including Project Background, Goals, Results, News, with options for which results to display, how to identify each member, and whether to show join links.
General Fund Status	Allows GA to accept donations to the study and to partially or fully sponsor a member through the project's General Fund.

process or the completion date for a kit that has already had its results reported. Note that some products such as the **YDNA12 + mtDNA** testing option involve more than one lab test. Each is tracked individually.

It is important to have results available in a format that is convenient for distribution to group members. You should also have your results available for display on group-related web sites.

The **Generate Y-DNA results for copy and paste** and the **Generate mtDNA results for copy and paste** links will provide you with your results in Excel, HTML, and pdf formats for these purposes. Each kit number is hotlinked to that member's personal page. If you copy and paste your results to another web site, make sure you remove these links from the table. If you do not do this, anyone who surfs into the website where your group results appear can link into the members' personal pages.

An interesting aspect of your group is the number of **Unique Haplotypes** it exhibits. If a large group has only a few unique haplotypes, its members are closely related. This could occur because of the way you found new

EXO-BIOLOGY AND HANDEDNESS

DNA is in the shape of a right-handed double helix, its sugar backbones coiling around each other much like the threads of a right-hand screw or the top of a jar. This handedness is very important in DNA's production of amino acids, the building blocks of the proteins that sustain life. Amino acids exist in both right handed (D-amino) and left handed (L-amino) mirror image forms, but right-handed DNA can only manufacture left-handed amino acids. As a consequence, life as we know it is exclusively based on left-handed amino acids.

There is no known reason why life insists on this asymmetry. But if life is present, so are left-handed amino acids.

But is the reverse true? Does an excess of left-handed amino acids indicate life?

In abiotic conditions, that is, where no life is present, L- and D-amino acids are present in a 50/50 mixture. For every L-amino acid, there is a D. Although an organism contains only the left-handed variety, when it dies, its amino acids will gradually convert to the 50/50 abiotic mixture of the two types.

Continued on p. 110

Exo-Biology (cont)

Carbonaceous meteorites are thought to be the remnants of dead comets that pre-date our solar system. They reveal information on the conditions that existed in the universe long before the earth was formed. Carbonaceous meteorites also contain amino acids, many of which are essential to life on earth.

In most of these meteorites, the concentration of left-handed and right-handed amino acids is the same. But in fragments of the Murcheson meteorite that fell in rural Australia in September 1969, there is an excess of some left-handed amino acids, in the range of 7 to 9%.[1] Is this a sign that life exists elsewhere in the cosmos? Did life begin on earth by an infection from bombardment by comets during the formation of the solar system? Or if life arose instead from abiotic conditions present in the early earth environment, how did life's preference for L-amino acids arise?

An important issue in this debate is contamination. While the presence of extremely rare amino acids in meteorites has firmly established their extraterrestrial origins, did the excess of L-amino acids

Continued on p. 111

members for your group. For example, if you have solicited only members at a family reunion, you might wind up with many genetically similar people. A large number of people sharing a small number of distinct haplotypes could also arise because your name is rare. In this case, anyone you find with your surname will have a common ancestor with the group unless he is the descendent of a nonpaternity event. In reviewing the number of unique haplotypes, it is a good idea to look at the differences that distinguish them. Having two haplotypes that differ by only one step and that are shared by a number of people has a different meaning in terms of your family history from having two groups of identical haplotypes that are widely separated. To learn more about the genetic diversity of different types of surnames, see the chapter *How Many Markers? How Many Participants?*

Family Tree DNA provides Group Administrators with **Family Project Website Setup** options. This makes it easy for you to setup the information displayed on and the functionality of your project website without a PhD in website design. The site allows you to explain and advertise your group through forms you can fill out showing your Project Goals, Project Background, Project Results, and

Project News. You can edit these whenever any of this information changes. Items that you leave blank will not appear on your website.

You have the option to automatically display the results appearing on your **Generate Y-DNA (mtDNA) results for copy and paste** pages. You may also choose to display a join link, your project surnames, a link for donations to your general fund, or a few words describing the fund. The URL for your website is chosen when you create the project. The URL is automatically inserted in the **Project Profile Page** at http://www.familytreedna.com/public/[THE_NAME_YOU_CHOSE] unless another web address is already referenced there. The GAP also includes the surnames you entered on your project page and provides a link leading to your website to allow you to view its contents.

As a Group Administrator, you will want to track the status of your group's slush fund. The **General Fund** link allows you to send a URL to your group members for donations, and provides instructions on how to use some of your slush to place an order. At the bottom of the page, there is a log of your General Fund activities, with a tally of your balance.

Exo-Biology (cont)

in the Murchison meteorite appear *before* or *after* the meteorite reached earth?

Although there is still debate on the issue, evidence indicates that the excess was there before the meteor arrived. Not only are the amino acids that show the L- excess rare on earth, there are four amino acids that are very common on earth whose L and D varieties appear in almost equal amounts. These facts minimize the possibility of terrestrial contamination from biological sources where only the L-form occurs.[2] Furthermore, individual amino-acids from Murchison are enriched relative to their terrestrial counterparts in the nitrogen isotope ^{15}N, confirming they are not of earthly origin.[3]

The earliest fossil record of life on earth shows a rich variety of organisms existing 3.8 billion years ago. There is no indication of life before this time. Is this because earlier records have been destroyed or that they are present but have not yet been found? Or is it because various primitive life forms hitched rides on comets and arrived about this time pre-made and thankful

Continued on p. 112

Exo-Biology (cont)

for an environment rich in nutrients after riding around in space cold and hungry for so many eons?

If we ever find a meteorite that exhibits *exclusively* left handed amino acids, that would be an almost unambiguous sign of life elsewhere in the cosmos. But if we ever find a meteorite that exhibits *exclusively right handed* amino acids, now that would be news….

1. Cronin, J. R., and Pizzarello, S., *Enantiomeric Excesses in Meteoritic Amino Acids*, Science, **275**, 951 (1997).
2. Kvenvolden, Keith A. et al, "Evidence for extraterrestrial amino-acids and hydrocarbons in the Murchison meteorite", Nature 228(**5275**), 923-6 (1970).
3. M. H. Engel and S. A. Macko, *Isotopic evidence for extraterrestrial non- racemic amino acids in the Murchison meteorite*, Nature 265 **389** (1997).

HOW MANY MARKERS?
HOW MANY PEOPLE?

GETTING STARTED

When you are considering joining a DNA study, one of the first questions you might ask is, "What can it tell me about my family?" The answer depends on how many people are in the study and how many markers are tested. The best way to show what you might learn from getting your DNA tested is with cladograms. A picture is worth a thousand words!

A cladogram (also called a haplodiagram or phylogenetic tree) gives a picture of family relationships. Cladograms are easier to interpret than tables of marker values. In this chapter, we use cladograms to explain various aspects of a DNA surname study, such as the benefits of testing on more markers and the influence that the number of members has on the conclusions that can be drawn about the surname. We also demonstrate that group results shown in a cladogram can give you information on the development of a surname. For example, a cladogram can indicate the relative ages of different branches of a family or the existence of multiple unrelated sources of a surname. We explain what pairwise mismatches are and how they can provide insights into finer points not obvious from a group's cladogram.

The reader is referred to other references for instructions on the how to construct cladograms and for information on the mathematical algorithms involved.[1] The cladograms presented in this chapter were constructed using the NETWORK software package offered by Fluxus Engineering.[2] It can be downloaded as freeware from www.fluxus-engineering.com.

ABOUT CLADOGRAMS AND PAIRWISE MISMATCHES

Figure 1 shows the cladogram created from the results of the Villareal surname study shown in Table 1. The haplotypes listed are those of six descendents of Diego Villareal (1601-1672), presumed to be the common ancestor of all Villareals from Texas and northern Mexico.[3]

In the diagram each open circle represents the haplotype of one or more members of the study, with the links between circles showing the markers where the haplotypes mismatch. The size of each circle is proportional to the number of people with the haplotype, and the lengths of the links are proportional to the number of mismatches between the haplotypes they join.

Three of the six Villareal haplotypes (members **100**, **400**, and **500**) are exact matches, represented by the larger open circle labeled **100**. Each of the remaining three haplotypes, depicted by the three small circles labeled **200**, **300** and **600**, differs by a single-step mismatch with respect to the main group. Numbers **300** and **600** each show different mutations at **DYS439** with respect to group **100** as indicated by the addition of different lower case letters after the name of the marker, **DYS439a** and **DYS439b**. Number **200** shows a mutation at **DYS464d**, denoted by **DYS464da**.

The first row in the table represents the assumed haplotype of the MRCA of the group members, also called the group's modal haplotype. It is composed of the most common value of each marker.

Figure 1 shows just one possibility for this group. The circles could have been shown with a common connection through **#600** rather than **#100**. But the computer algorithms used to construct a cladogram produce trees with the shortest and least number of connections. These are called maximum parsimony (MP) trees. In this case

Table 1. Villareal haplotypes.[3] The first row is assumed to represent the haplotype of their common ancestor Diego Villareal.

DYS	393	390	19	391	385a	385b	426	388	439	3891	392	389	458	459a	459b	455	454	447	437	448	449	464a	464b	464c	464d
MRCA	13	25	13	9	18	18	11	12	12	13	12	30	15	9	9	11	12	26	14	20	32	15	16	16	18
100	13	25	13	9	18	18	11	12	12	13	12	30	15	9	9	11	12	26	14	20	32	15	16	16	18
200	13	25	13	9	18	18	11	12	12	13	12	30	15	9	9	11	12	26	14	20	32	15	16	16	19
300	13	25	13	9	18	18	11	12	13	13	12	30	15	9	9	11	12	26	14	20	32	15	16	16	18
400	13	25	13	9	18	18	11	12	12	13	12	30	15	9	9	11	12	26	14	20	32	15	16	16	18
500	13	25	13	9	18	18	11	12	12	13	12	30	15	9	9	11	12	26	14	20	32	15	16	16	18
600	13	25	13	9	18	18	11	12	11	13	12	30	15	9	9	11	12	26	14	20	32	15	16	16	18

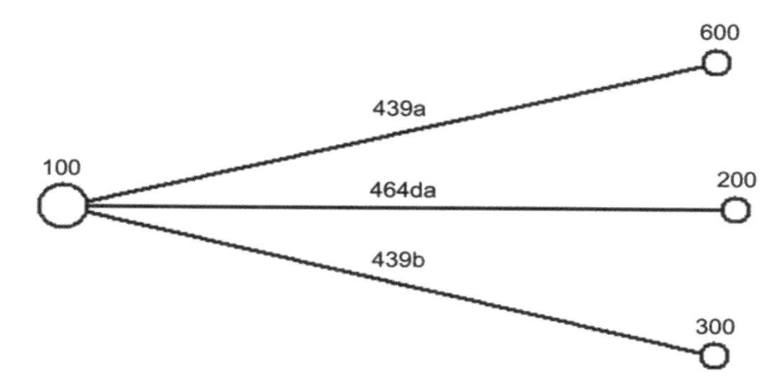

Figure 1. Simple cladogram created from the results in Table 1 of the Villareal surname study.

there is only one possible diagram with maximum parsimony, but for larger study groups there is usually more than one. In the following discussion, we show only one MP tree for every study group we use as an example.

A graph of the pairwise mismatches for this group of Villareals is shown in Figure 2. Each member's haplotype has been compared to every other member's haplotype and the number of mismatches noted. The graph indicates there are three pairs of haplotypes that match exactly, there are nine pairs that show a single step mismatch, and there are three pairs that show two mismatches. See Table 2.

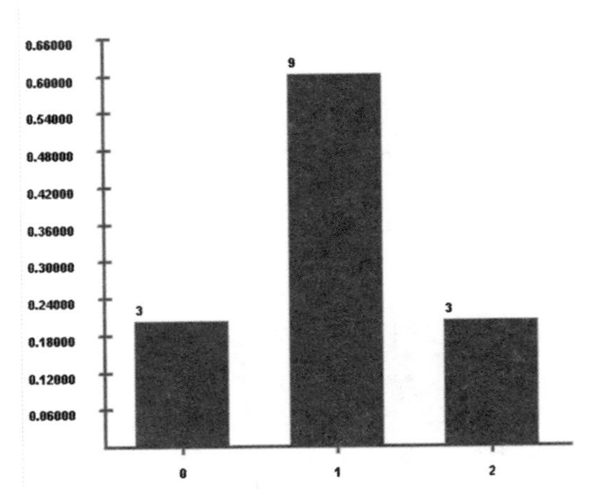

Figure 2. Pairwise mismatches of the Villareals in Table 1.

THE MUMMA STUDY

As of the summer of 2005, the Mumma surname study had 68 members from the U.S., Germany, and Sweden who had been tested. Variations on the name that have been included are Muma, Mummau, Maummah, Mumaugh, Moomaw, Moomau, Moomey, Mumme, Mummey, and Reenstjerna.

The Mumma surname is not very common, so it probably had only a few origins. Because many American and European members have reliable documentation linking them to their Mumma ancestors, the study has been able to identify specific haplotypes for one of the American branches, to link all American branches with their European roots, and to identify which variations on the surname spelling have a genetic link to the group. It offers an opportunity for Mummas who do not know their ancestry to identify which family lines they belong to.[4]

Table 2. Villareal pairwise mismatches.

No. of Mismatches	No. of Pairs	Haplotypes	Mutations
0	3	100 & 400	---
		100 & 500	---
		400 & 500	---
1	9	100 & 200	464da
		100 & 300	439b
		100 & 600	439a
		400 & 200	464da
		400 & 300	439b
		400 & 600	439a
		500 & 200	464da
		500 & 300	439b
		500 & 600	439a
2	3	200 & 300	464da, 439b
		200 & 600	464da, 439a
		300 & 600	439a, 439b

According to the Mumma web site at www.mumma.org, most of the Mummas in the U.S. are descended from three immigrants who arrived in America in the early to mid 1700s: Jacob (arrived 1720), Leonard (arrived 1721), and Peter (arrived 1748). Thanks to the DNA study, it has been shown that these three Mumma immigrants were related (although they were probably not brothers) and that their American descendents are genetically linked to Mummas living in Germany and Sweden who are descended from William Mumma (b. 1543). Some members of the study with variations on the surname were ruled out as being members of this Mumma family, while others with different surnames (Bell, Reenstjerna) were ruled in as belonging to this family.

One exciting discovery is that all descendants of Leonard Mumma have a different value for the marker **DYS570** from all the other Mummas. While the rest of the Mummas show a value of **DYS570** = 17, Leonard's descendents show **DYS570** = 16.

Because descendents of each of Leonard's sons have this mutation, Leonard himself must have carried it and the mutation can be used to uniquely identify Mummas that belong to Leonard's branch of the family. Testing the **DYS570** marker is important in sorting out Mummas with ambiguous written genealogies and assigning family lines to Mummas who do not know their ancestry.

The four progenitors of the Mumma family along with their dates of birth or immigration are shown in Table 3. Also listed are the largest clusters of identical haplotypes associated with each progenitor. This information will be useful in understanding how testing on more markers can help to tease apart different family lines.

MORE MARKERS, MORE INFORMATION

The greater the number of markers two people are tested on, the better defined their relationship is. Testing on 20 markers or more will give a reasonably accurate estimate of the time back to their most recent common ancestor (MRCA).[5] Using less than 20 will not give a useful estimate, but can be valuable in ruling out a relationship or identifying a nonpaternity event. This is useful for people with a rare surname.

Cladograms are an excellent way to illustrate this. As more markers are included, the branches of the group's cladogram grow longer and "spread out" so that different branches of a family are better defined.

Table 3. Largest groups of identical haplotypes in the Mumma study and their progenitors.

Progenitor	Date	Group
William	1530	M-10
Leonard	1721	M-3
		1861
		M-19
Jacob	1720	1946
		615
Peter	1748	7057

12-Marker Tests

Figure 3 shows a cladogram of the results of testing Mummas on 12 markers. All but one of them are related within two mutations of each other. The largest cluster of 30 members (**M-10**) includes descendents of William (**W**), Leonard (**L**), and Jacob (**J**) Mumma. The cluster is one mismatch away from three singletons and from three smaller clusters of nine, six, and five haplotypes. Only Peter's (**P**) line (**#7057**) has been clearly differentiated by the 12-marker test. With only 12-marker test results we can conclude that 30 of the Mumma descendents are related in the recent past. This lower power test is unable to resolve three out of four of the Mumma lineages.

Even on this 12-marker test, there is a study member who lies so far from the main group that he cannot be conveniently included in the figure. An "outlier," **#5949** is separated from his nearest neighbor by 12 mutations. (Some are multiple mutations on the same marker). He must be the descendent of a nonpaternity event that occurred through an adoption, a name change, or an illegitimacy. Its remote location in the cladogram illustrates that even a 12-marker test can identify outliers who do not have a genetic relationship with the rest of the group.

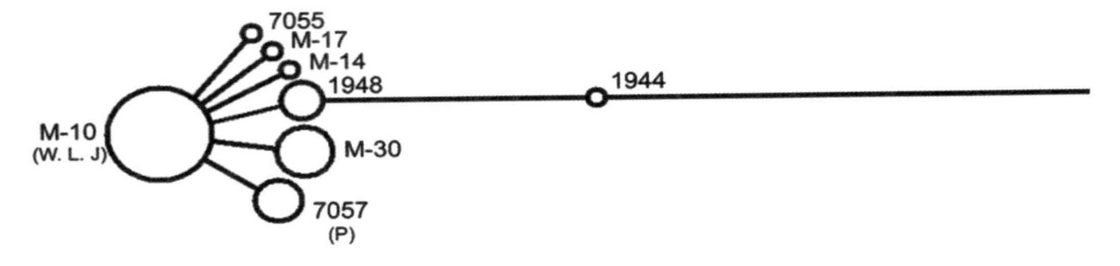

Figure 3. Tests on only 12-markers can identify a non-paternity event, but cannot resolve different family lines. W = William, L = Leonard, J = Jacob, P = Peter Mumma. Each cluster is labeled with its constituent groups.

In the chapter *Are We Really Cousins?* we define study members as closely related if there is at least a 50% probability they share a common ancestor who lived within the last 200 years, and as distantly related if he lived within 500 years. The number of mismatches they can have and still be considered close or distant relations depends on the number of markers tested, which markers are tested, and the estimated mutation rate of each marker. Table 4 shows the maximum number of mismatches for closely and distantly related haplotypes for 12, 25, and 37 markers and the assumed average mutation rates.

We are using average mutation rates for our calculations because the individual mutation rates used by genetic genealogy testing companies are considered proprietary. The same number of mismatches may not yield exactly the same probability of a relationship as the online FTDNATiP™ Time Predictor Calculator since it uses individual mutation rates. The online calculator is more accurate.

Keep in mind that all results must be interpreted with the word "probably" implicitly attached. Occasionally a father-son pair will differ on one of 12 markers although they are obviously closely related. In the overwhelming majority of cases, however, they will match exactly. If two people differ on one of twelve markers, unless a paper trail indicates otherwise, we can conclude that they are "probably" not closely related.

Based on the average mutation rate for a 12-marker test, closely related haplotypes can differ by no more than 0.08 mismatches, that is, they must be identical. Figure 4 shows this as a circle including only the **M-10** haplotype. This main group is used for this example, but any of the other haplotypes or haplotype groups could be used as the center.

Table 4. Maximm number of mismatches for haplotypes to have a 50% chance of being related in the last 200 or 500 years.

N = number of markers	12	25	37
R = average mutation rate	0.0039	0.0044	0.0053
k = mismatches (200 yr.)	0.08	1.25	2.42
k = mismatches (500 yr.)	1.21	4.14	7.14

Distantly related haplotypes that lie within 1.21 mismatches are included in a larger dashed circle centered on the **M-10** group in Figure 4. **#1944** lies far outside of the larger circle, so it is unlikely that he is related to the main group **M-10**, but he appears within 2 markers of the group represented by **#1948**. There is a small chance that **#1944** may show a relationship when more markers are included in the test, so we won't label him as an outlier yet.

Figure 5 shows a graph of pairwise mismatches for a 12-marker test. The large peak near the origin shows that most pairs are close matches. The peak is relatively narrow at four mismatches wide. Fifty percent of the pairs show one mismatch. The small peak to the far right is caused by the outlier **#5939** who mismatches the rest of the group on a large number of markers.

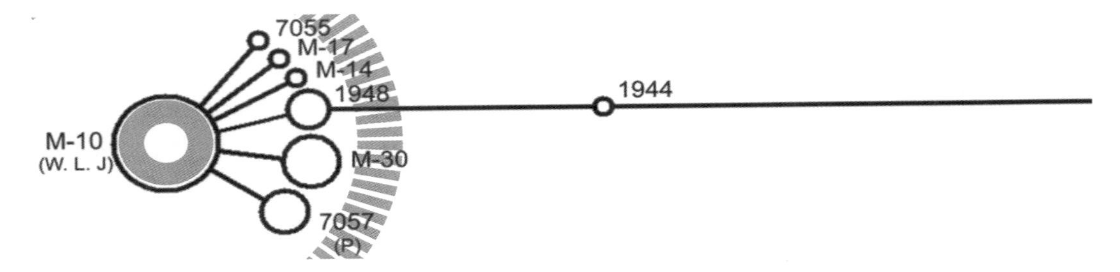

Figure 4. Haplotypes in the solid circle have a 50% chance of being related to M-10 within 200 hundred years; those in the dashed circle, within five hundred years.

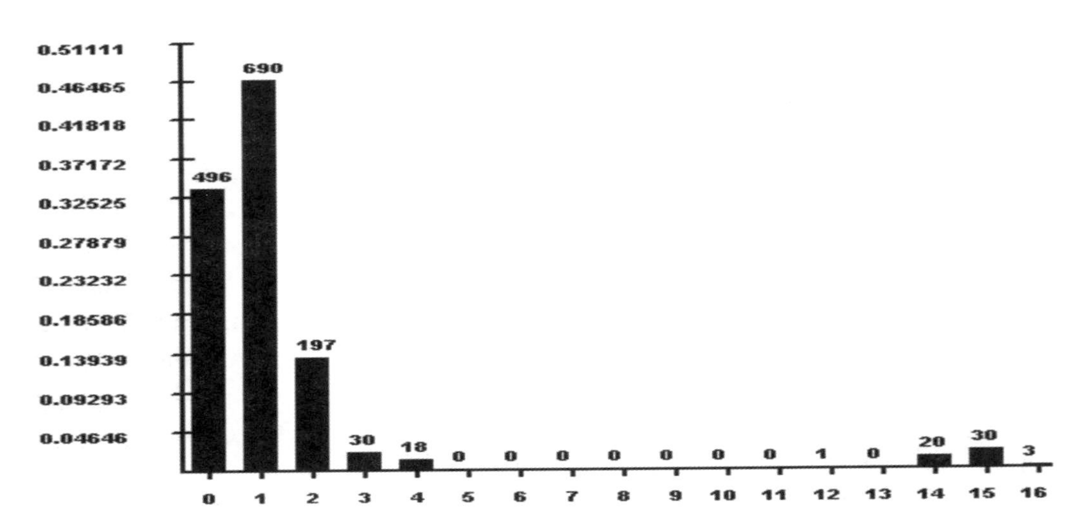

Figure 5. Mumma pairwise missmatches for the 12-marker test.

25-Marker Tests

Testing a group on more markers results in a more complex network of relationships. See the cladogram of the results of testing Mummas on 25 markers in Figure 6. (**#5949** has been eliminated as an outlier in the cladogram, but has been retained in the graph of pairwise mismatches.)

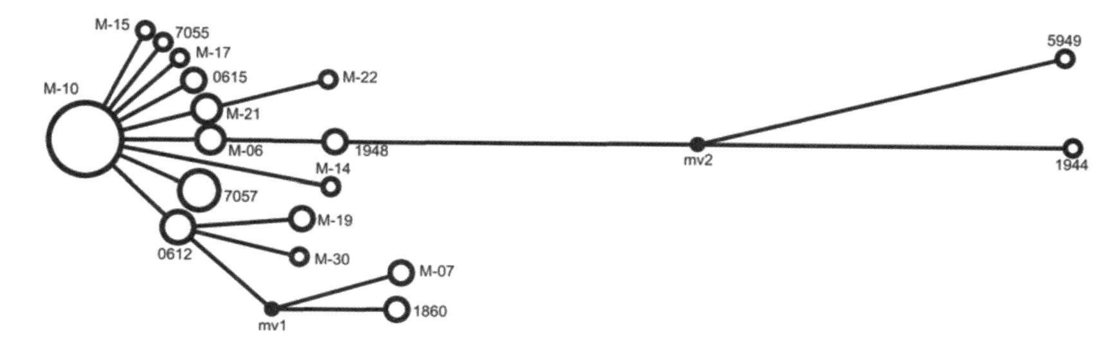

Figure 6. Mumma results for the 25-marker test.

Some haplotypes now show more mismatches, so that they have either split off from the group they previously belonged to, or they have moved farther away from their nearest neighbor. For example, the core group still represents haplotypes from three out of four family lines, but now contains only 21 members. Nine people from Leonard's and Jacob's lines who used to be in this group now appear in satellite clusters. Cluster **#1948** formerly identified with Leonard and Peter (Figure 3) has also fragmented into two smaller groups. (Peter's cluster **#7057** of six members remains intact). Although adding 13 markers better distinguishes the members of the various family lines, there is still ambiguity in assigning several group members to a specific family branch.

Originally **#1944** showed only two mutations with respect to his nearest neighbor on the 12-marker test, but is now separated by nine. He is confirmed as a nonpaternity event.

To calculate this more complex network, the program has introduced *median vectors* **mv1** and **mv2** shown as a small closed circles. **mv1** is towards the bottom of Figure 6. A median vector represents a hypothetical haplotype that might belong to a future member of the group or one that has died out because a Mumma ancestor had no sons or only sons with mutations. In any analysis, median vectors should be treated the same as haplotypes that are already in the group. Because it is only one mutation away from a cluster with both Leonard's and Jacob's descendents in it, the **mv1** haplotype probably belongs to one of these branches of the family.

For a 25-marker test, closely and distantly related haplotypes are considered to be separated by no more than 1.25 and 4.14 mismatches, respectively. See Table 4. Two circles have been added to Figure 6 to define close and distant neighborhoods around **M-10**. (Figure 7).

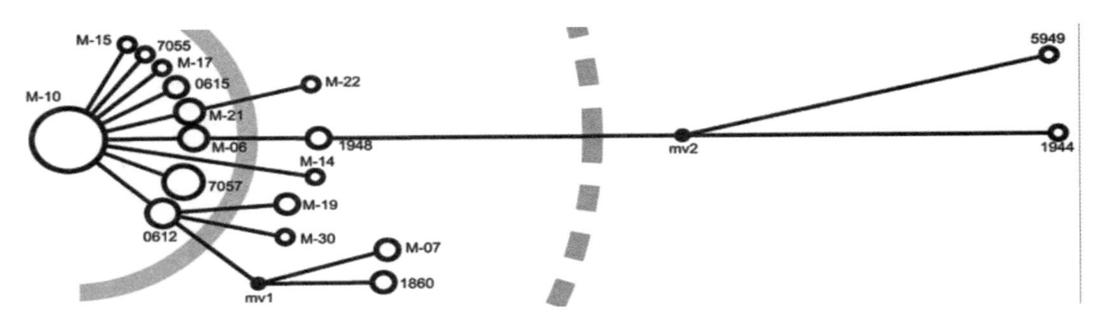

Figure 7. Haplotypes that fall within the solid circle have a 50% chance of being related to M-10 in last 200 yrs.

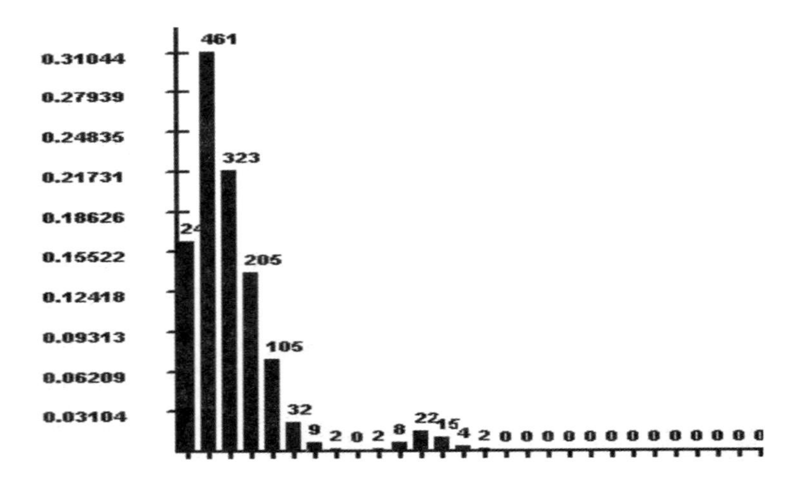

Figure 8. Pairwise mismatches for the Mumma surname study for a 25 marker test, including outliers.

The graph of pairwise mismatches for the 25-marker results is shown in Figure 8. Most pairs still mismatch on a low number of markers, but the peak near the origin is now wider and shorter, reflecting the increased number of mismatches that occur because of the increased number of markers tested. The small peak (**#5949**) at the far right has moved farther away from the main group for the same reason. The 25-marker results also show a second small peak centered at 11 mismatches. This is due to the increased genetic distance of **#1944** from the rest of the group, as shown in the cladogram.

Testing on more markers allows more mismatches for the same degree of closeness. Whereas testing on 12 markers requires an exact match for two people to be considered closely related, testing on 25 markers opens this definition up to include study members who show 0.67 mismatches. A 12-marker test allows for at most 1.21 mismatches to meet the requirements for a distant relationship, but a 25-marker test allows 4.14 mismatches.

<u>37-Marker Tests</u>

The group has an even more complex structure when tested on 37 markers because of the appearance of new mismatches. (See Figure 9). The large cluster of haplotypes that used to be at the center of the cladogram has now broken into two fragments with different haplotypes. The new core cluster (**M-10**) is the smaller of the two containing the haplotype of the presumed MRCA of the whole family. For the Mumma 37-marker test, the peak representing closely matching pairs is still the largest, but is more spread out. (See Figure 10). It is now wide enough to crowd the smaller peak caused by outlier **#1944**. The peak on the far right associated with **#5949** has moved even farther away.

When considering whether a study member is an outlier, it is important to look at the whole network, and not judge him based only on his distance from the core group. In the Mumma study, some haplotypes definitely belong to Leonard's branch of the family even though they lie far away from the core group. Although these haplotypes could be mistakened for outliers, they are closely related to intermediate haplotypes that belong to Leonard's branch, and these intermediate haplotypes are closely related

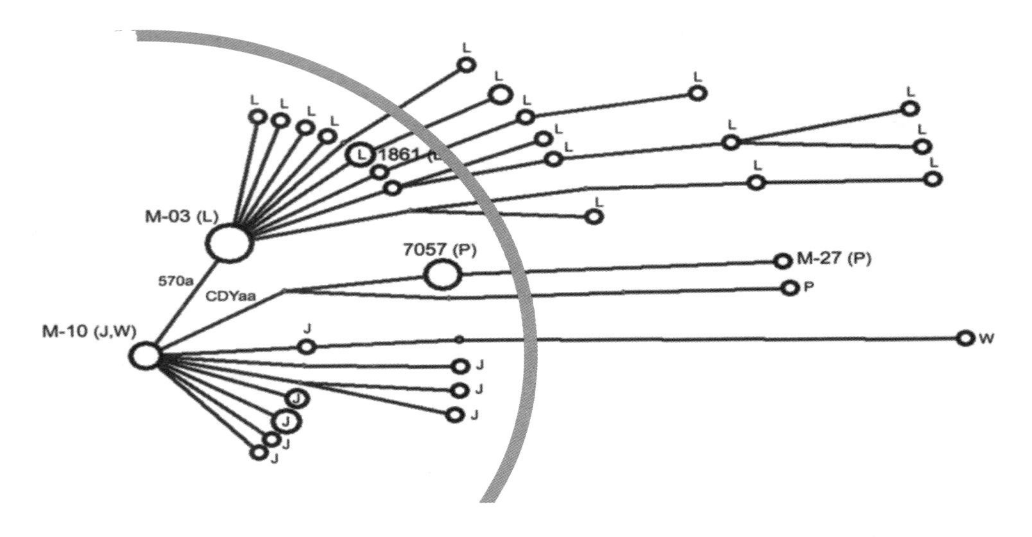

Figure 9. Haplotypes that fall within the circle have at least a 50% chance of being related to M-10 up to two hundred years ago; all others have more than a 50% chance of being related to M-10 within the last five hundred years.

to the core group. Though they fall outside of our definition of a distant relationship with the core group, they are part of a network of haplotypes belonging to Leonard's family line. They should not be eliminated from the study as outliers.

Table 5 gives a summary of how the groups of haplotypes that were identical on lower power testing are distinguished by testing on more markers. Note that with 12 markers, we were able to spot nonpaternity events, but to be related in the recent past (within the past 200 years) required two people to match identically. On the 12-marker test, only one of the Mumma lines was distinct from the others. Testing on 25-markers showed a more complicated network of relationships. Some haplotypes showed more mismatches so that the various family lines became better separated. However, there was still uncertainty in assigning several members to a specific branch. The lineages were finally resolved by the 37-marker test. As more markers were added, the Mumma haplotypes spread out and their cladogram took the shape of a double star, with the group's modal haplotype located at the center of one of the stars.

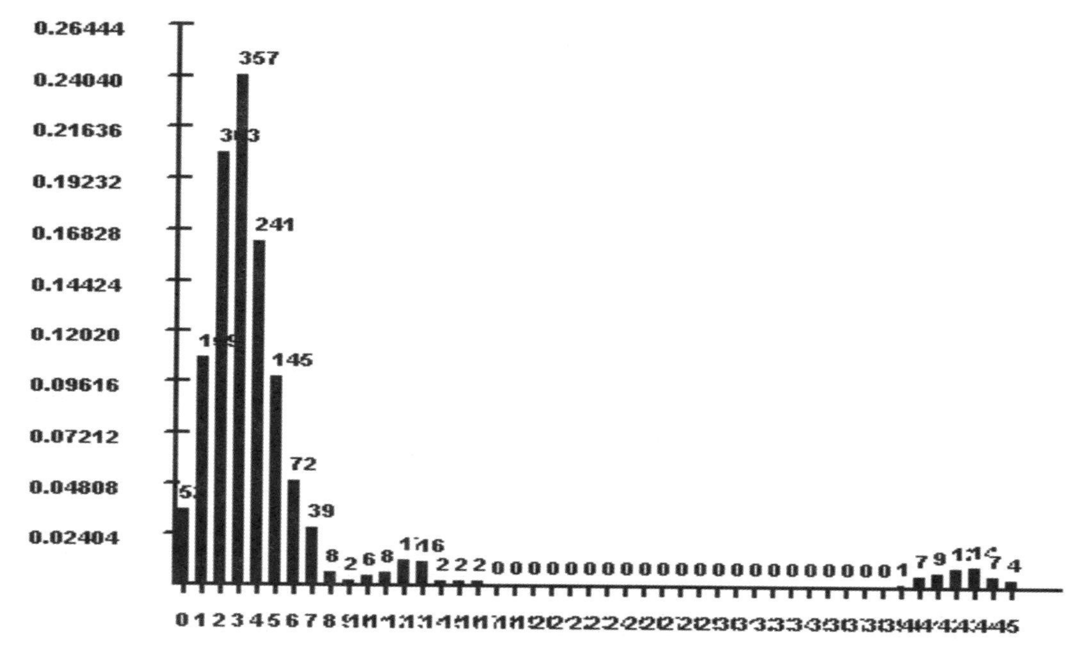

Figure 10. Pairwise mismatches for the Mumma 37-marker test.

REFERENCES–HOW MANY MARKERS? HOW MANY PEOPLE?

1. C. Fitzpatrick, *The DNA Detective, Forensic Genealogy*, Rice Book Press, 2005.
2. In all examples, the data have been corrected for the overlap of markers DYS389I and DYS389II, that is the values of DYS389a = DYS389I and DYS389b = DYS389II – DYS389I have been used. A second correction has been made for the hypervariable DYS464a, b, c, and d markers, where they have been assumed to mutate according to the infinite allele model. Differences on this multiallele marker have been taken as either a 0 or a 1, that is, they are considered either mutated or unmutated and the difference between the actual marker values is not taken into account.
3. www.mumma.org
4. http://www.mumma.org/DNA.htm.
5. Walsh, "Estimating the Time to the Most Recent Common Ancestor for the Y-Chromosome of Mitochondrial DNA for a Pair of Individuals', *Genetics*, **158**, 897-912 (June 2001).

Table 5. Tests on more markers can tease apart family lines.

<table>
<tr><th>12-Marker Groups</th><th>25-Marker Groups</th><th>37-Marker Groups</th><th>Progenitor</th></tr>
<tr><td rowspan="12">M-10 (30)</td><td rowspan="4">M-10 (21)</td><td>M-10 (4)</td><td>William</td></tr>
<tr><td>M-03 (8)</td><td>Leonard</td></tr>
<tr><td>1946 (3)</td><td>Jacob</td></tr>
<tr><td>6 Singletons</td><td>Leonard or Jacob</td></tr>
<tr><td rowspan="2">M-07 (2)</td><td>M-07 (1)</td><td>Jacob</td></tr>
<tr><td>M-09 (1)</td><td>Leonard</td></tr>
<tr><td>0615 (2)</td><td>0615 (2)</td><td>Jacob</td></tr>
<tr><td rowspan="3">M-21 (3)</td><td>M21 (1)</td><td>Leonard</td></tr>
<tr><td>975 (1)</td><td>Leonard</td></tr>
<tr><td>3314 (1)</td><td>Leonard</td></tr>
<tr><td>M-22 (1)</td><td>M-22 (1)</td><td>Leonard</td></tr>
<tr><td>M-15 (1)</td><td>M-15 (1)</td><td>Leonard</td></tr>
<tr><td rowspan="6">M-30 (9)</td><td>M-30 (1)</td><td>M-30 (1)</td><td>William</td></tr>
<tr><td rowspan="2">0612 (4)</td><td>0612 (1)</td><td>Jacob</td></tr>
<tr><td>1860 (3)</td><td>Leonard</td></tr>
<tr><td>M-19 (2)</td><td>M-35 (2)</td><td>Leonard</td></tr>
<tr><td rowspan="2">1860 (2)</td><td>1860 (1)</td><td>Leonard</td></tr>
<tr><td>M-35 (1)</td><td>Leonard</td></tr>
<tr><td rowspan="2">7057 (6)</td><td rowspan="2">7057 (6)</td><td>7057 (5)</td><td>Peter</td></tr>
<tr><td>M-27 (1)</td><td>Peter</td></tr>
<tr><td rowspan="5">1948 (5)</td><td rowspan="2">1948 (2)</td><td>1948 (1)</td><td>Peter</td></tr>
<tr><td>M-16 (1)</td><td>Leonard</td></tr>
<tr><td rowspan="3">M-6 (3)</td><td>M0-6 (1)</td><td>Leonard</td></tr>
<tr><td>M-11 (1)</td><td>Leonard</td></tr>
<tr><td>3317 (1)</td><td>Leonard</td></tr>
<tr><td>M-14 (1)</td><td>M-14 (1)</td><td>M-14 (1)</td><td>Leonard</td></tr>
<tr><td>M-17 (1)</td><td>M-17 (1)</td><td>M-17 (1)</td><td>Leonard</td></tr>
<tr><td>7055 (1)</td><td>7055 (1)</td><td>7055 (1)</td><td>Leonard</td></tr>
</table>

WHAT CAN DNA TELL ME ABOUT MY SURNAME?

Until relatively recently, people went by a single name. Surnames were not needed. A person only had a small circle of family and friends to worry about. If there was a need to distinguish between two people with the same given name, a by-name was appended for this purpose, such as "John who lives across the fields," or "John with the big head." As population increased, this practice became more common, with add-ons eventually becoming hereditary surnames in Western Europe between the twelfth and sixteenth centuries.[1] Surnames are usually derived from one of four sources:

- a geographical location
  - Netherwood–someone who lives in the lower woods
  - Banker–someone who lives near the bank of a river
  - Atwood–from the Saxon "atte Wood" or at the wood

- an occupation
  - Baker, Baxter (male and female bakers)
  - Fawcett (judge)
  - Latimer (translator)
  - Smith
- a physical attribute
  - Kennedy (from the Gaelic *ceannaideach* or "ugly head")
  - Armstrong
  - Beard

- a patronymic
  - Anderson, Andersen
  - McDonnell, O'Donnell
  - Fitzpatrick

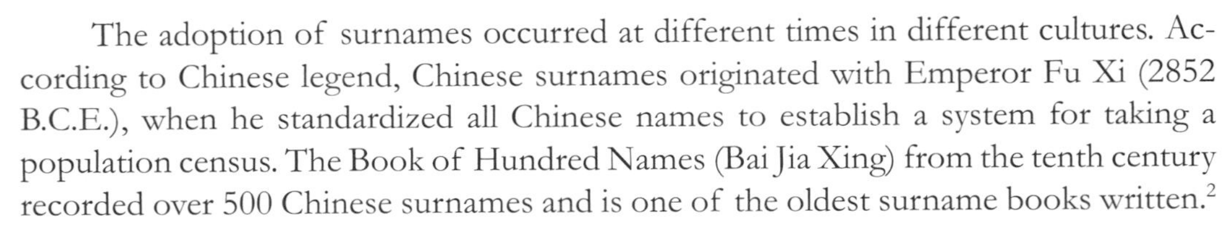

The adoption of surnames occurred at different times in different cultures. According to Chinese legend, Chinese surnames originated with Emperor Fu Xi (2852 B.C.E.), when he standardized all Chinese names to establish a system for taking a population census. The Book of Hundred Names (Bai Jia Xing) from the tenth century recorded over 500 Chinese surnames and is one of the oldest surname books written.[2]

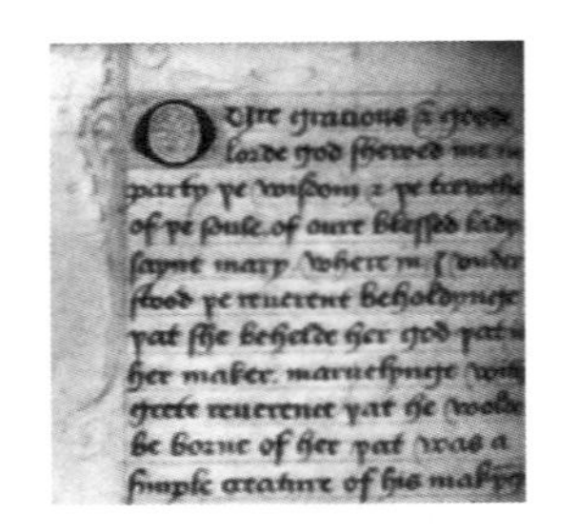

Oure gracious & goode
lorde god shewed me
party ye wisdom & ye trewthe
of ye soule of oure blessed lady
saynt mary where in I under
stood ye reuerent beholdynge
yat she behelde her god yat
her maker marvelynge with
grete reuerence yat he wolde
be borne of her yat was a
simple creature of his makyng

The ancient Romans had three names, a forename (*praenomen*), of which there were fewer than 20; the tribe name (*nomen*); and finally the family name (*cognomen*); e.g., Caius Julius Caesar, or Caius of the Caesar family of the Julian tribe. Sometimes an additional name was included (*agnomen*) as a nickname or honorific such as Africanus, for victory in Africa, in the case of Scipio.[3] But with the fall of the Roman Empire, surname usage waned, and by the Middle Ages most Europeans were known by only a single given name.

Probably the earliest recorded fixed European surname in modern times is O'Clery (Ó Cleirigh), as noted by the Irish *Annals of the Four Masters*, compiled in the 17th century in County Donegal, Ireland by Brother Michael O'Clery. The *Annals* record the death of Tigherneach Ua Cleirigh, lord of Aidhne in County Galway in the year 916.[4] The Domesday Book (1085–1086) compiled by William the Conqueror recorded some family names, but they did not become common until the 1200s,[5] when they had become a virtual necessity.

According to Brian M. Scott, a member of the Academy of St. Gabriel, a group that researches names and armory,[6] in the thirteenth century most men and women in England shared the same few given names. His compilation of names drawn from the custumals of 21 manors held by the Abbey of Bec[7] showed that one of every seven men was named William and that the five most common names (William, Richard, John, Robert, and Hugo) were used by 45% of the male population. The ten most popular male names accounted for 70% of the men listed. Women's names were just as common. The most frequent name Matilda and its variants were used by one of every six woman, with the top five names (Matilda, Alice, Agnes, Edith, Emma) representing just over half the female population. The top 12 names accounted for about 70% of the women listed.

THE HEAD TURK

There is at least one surname that died out because it was possessed by only one person who had no male children. In 1932 the Turkish nation became the last in Europe to adopt family names. They were usually chosen through personal preference based on a family characteristic or as a matter of personal pride. Kemel Ataturk was the leader of Turkey who forced this change on the Turkish people in his drive to bring the country into the modern era. The surname he chose for himself, "Ataturk", means "Father of the Turkish Nation". Since he had no male children and was such a hard act to follow, it is appropriate that the world will only know one Ataturk. The Ataturk surname cladogram looks like "●".

As population increased, surnames became critical in distinguishing among people by the same given name. Even so, surnames were often depicted graphically due to the high rate of illiteracy. Although the introduction of the printing press to England by William Caxton[8] in 1476 helped to improve the level of literacy and to standardize English spelling and grammar, variations in the spelling of surnames remained. Around the time of his birth in April 1564, William Shakespeare could have found 27 different variations in the spelling of his surname, including Shagspeare and Chacsper.[9]

Other cultures adopted surnames on their own schedules. Surname usage was not enforced among Filipinos until 1849,[10] and surnames were made a legal requirement for Turks[11] only in 1924. Today there is a wide variety of surname styles, from Iceland, where the surname is a patronymic that changes from generation to generation (Hermannsson, Oddsson), to Spain where the surname is often a combination of the mother and father's surnames (Garcia y Lopez), to Eastern Europe and the Orient where the family name comes before the given name.

Characteristics of Family Names

So where does *your* surname fit in? Is it common or rare? How far back in time did your family use your surname? Are there spelling variations of your name? If so, are you genetically related to people with these variations? A cladogram of the results of your DNA name study can help answer these questions.

General Features

Seeing Stars. As an ancestor's haplotype is handed down in the family, it will mutate randomly, causing different genetic branches. After generations, many of the ancestor's descendants will probably still have his unmutated haplotype, with a few who exhibit a small number of mismatches. The group's cladogram will be a single star-like structure. Since the haplotype of the common ancestor will have been in existence longer than those of any of his mutated descendents, for well-sampled family groups, the haplotype of the MRCA will probably be located at the center of the star and will be that of the largest group in the study. The mutated haplotypes will appear either as singletons or as smaller groups that radiate from the center through one or two mis-

Figure 1. The cladogram of the Allred family forms a star pattern. This indicates that the name was in use for a long time.

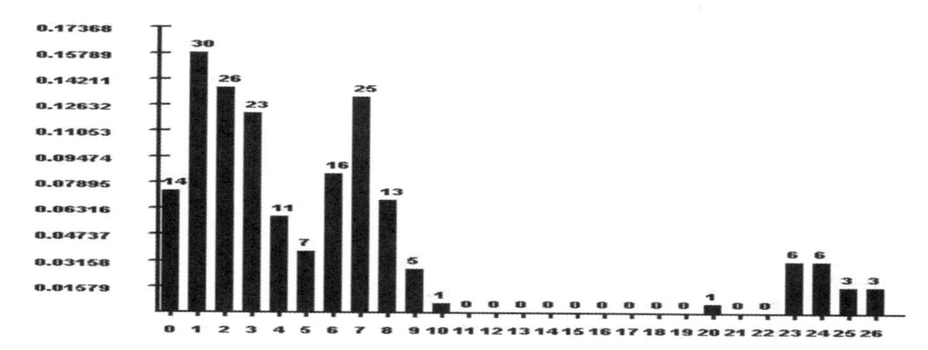

Figure 2. Pairwise mismatches for the Allred surname study including outliers.

matches. (See Figure 1 for the cladogram of the Allred surname study.) As more people join the study who have this common ancestor, they will be added to the star, either along new rays, along already existing rays, or as part of the star's center. If a surname has had more than one origin, multiple separated stars appear in a single cladogram.

The close relationship of most members of the group is reflected in its graph of pairwise mismatches. Most pairs of group will mismatch on a low number of markers. A group member who is the product of a nonpaternity event will mismatch most other group members on a large number of markers. Nonpaternity events will typically appear as pairwise mismatch peaks well-separated from the main group. Figure 2 shows the pairwise mismatches for the Allred study corresponding to the cladogram in Figure 1.

The MRCA of a group exhibiting such a star pattern was not only the genetic ancestor of the group, he used the name. This common ancestor could have been the first one to have used the name. Or he could have lived through a population bottleneck as the only male survivor of a family who died due to war or disease, or after the rest of the family changed their surname for political reasons. He could also have been the only male offspring of a family.

If the population with a surname is not well-sampled by its DNA study, the group can be "lopsided," favoring some branches of the family over others. In this case, the cladogram will not give a true picture of the surname structure. But as the study grows in size and has a more evenly distributed sample, missing branches of the family will appear, more realistically showing the surname structure. It may not be a star at all.

Very old surnames may form a double-star pattern in a cladogram, where two stars are separated by only one or two mutations. A surname can have this type of structure if it was adopted by two close relatives at about the same time in the distant past, or if enough time has gone by to allow the haplotype of a descendant of the family's progenitor to accumulate its own distinct sets of mutations. In this case, the time period of the MRCA of the combined group can be calculated based on the average number of mismatches of all group members. The cladogram of the 37-marker results of the Mumma surname study shows this type of double star pattern, with the centers of the two stars represented by groups **M-10** and **M-03**. See Figure 9 in the chapter *How Many Markers? How Many People?*.

In associating the MRCA at the center of a star pattern with the progenitor of a family line as indicated by paper documentation, there is one thing you should keep in mind. Even though there might be reliable records documenting this common ancestor, you cannot say for sure who he really was. For example, about 25% of the members of the Fitzpatrick DNA study form a star pattern. Some of the members of this star trace their ancestry, at least on paper, to Bryan McGiollaphadraig, born 1485, Lord and First Baron of Upper Ossory, now part of County Laois, Ireland. However, DNA analysis cannot confirm that their common ancestor was Bryan himself. The common ancestor could have been a cook in Bryan's household who just got lucky. It is impossible to know for sure since we do not have a DNA sample from Bryan himself (nor from his cook).

Outliers and Nonpaternity Events. While the appearance of an outlier in a group hints that there was a nonpaternity event in the family, the absence of an outlier does not rule it out. If a family has occupied the same geographical area for many generations, it is highly likely that if a wife has had an affair, it was with a relative of her husband. If a liaison was with the brother of her husband, a resulting nonpaternity event would be genetically undetectable. If the nonpaternity event was the product of a liaison with her husband's distant cousin, the child might show enough mismatches to create suspicion, but there would be no proof of an illegitimacy because of the statistical nature of mutations. We'd like to provide an example of this, but nobody answered our ad in the paper.

Three Kinds of Surnames

Rare Surnames. There are many ways that a rare surname can come into use, but it usually has had only one source, for example a specific (rather than generic) geographical location, a personal nickname, or even a mistake. Over the generations, mutations will have accumulated in the family haplotype creating different branches of the family, so that the structure of a rare surname will take the shape of a star pattern, and nonpaternity events will not be difficult to detect. Mumma and Allred are good examples of rare surnames showing this type of structure. (See Figures 9 and 10 in the chapter *How Many Markers? How Many People?* for the cladogram and graph of pairwise mismatches for the Mummas, and Figures 1 and 2 above for the Allreds.)

Common Names with Multiple Sources. Names derived from professions or geographical locations are very common, many originating in different countries during different periods of time. As time passes, and populations are displaced due to wars, famines, and other world events, people have relocated and often converted their surnames to a different language, so that the geographical origins of their names may be long forgotten. Because of the diversity in its origins, this kind of surname can include several haplogroups.

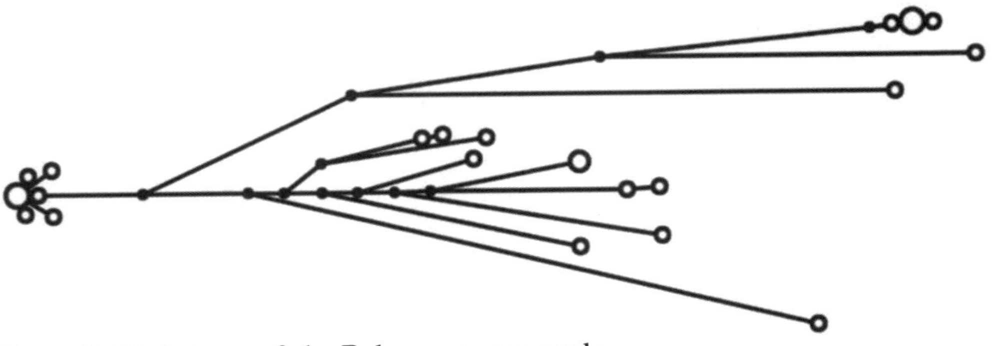

Figure 3. Cladogram of the Baker surname study.

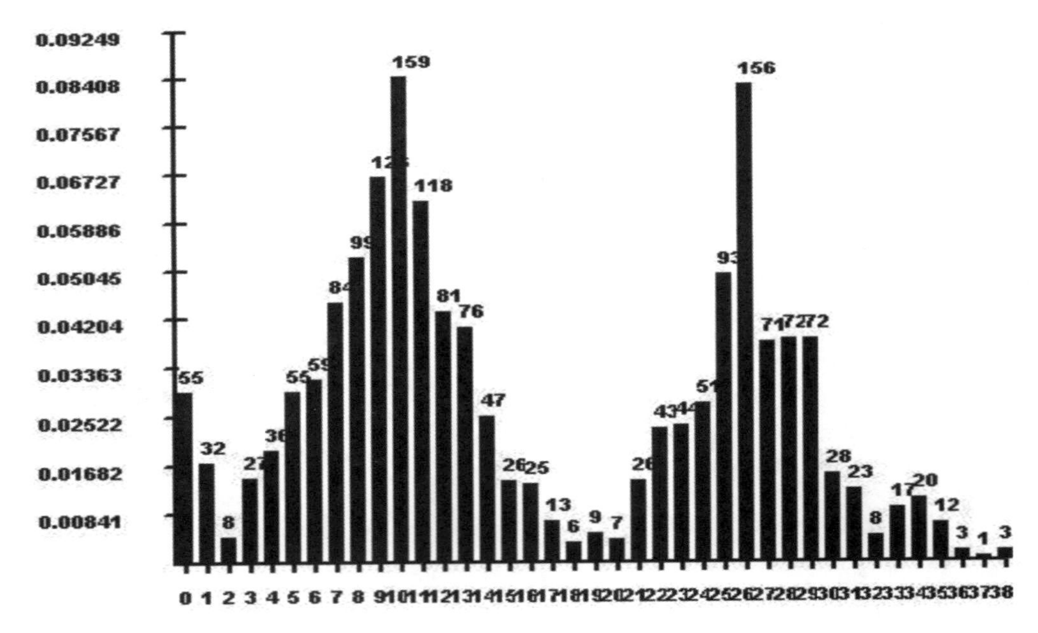

Figure 4. Pairwise mismatches for the Baker surname including outliers.

The genetic network of a common surname will probably show many well-defined clusters representing older branches of the family, interspersed with singletons representing newer branches, nonpaternity events, or incomplete sampling. There will be several well-separated clusters of similar haplotypes in the group. This will show up in the graph of pairwise mismatches as a peak near the origin representing pairs of study members from the same cluster, and one or more peaks farther away from the origin representing mismatches between pairs of people from different clusters. For an example of this kind of surname structure, see Figures 3 and 4 for the cladogram and graph of pairwise mismatches for the Baker surname study.

Prosperity, Persecution and Spaghetti. Some surnames do not fit into either of these categories. These surnames have been in use for a long time, but they also have been adopted by a large number of families in the recent past, usually in the same geographical region, perhaps for economic or political reasons. Unlike a common name with many origins, this type of surname shows one or at most two large clusters of closely related haplotypes, surrounded by a genetic spaghetti of singletons and very small groups. Most of the members of the study will belong to the same haplogroup. Such a surname structure is typical of the Irish Clans.

Assuming the study includes a well-sampled population, the appearance of a large cluster, perhaps in the shape of a star pattern, represents an older, dominant branch of the family. Since wealthy families adopted surnames centuries before common folk due to the need to establish inheritance, the progenitors of these older core groups were probably powerful and influential in their geographic areas.[12]

The appearance of many unrelated haplotypes indicates that the surname was adopted relatively recently by a large number of loosely related families. This could have been due to a time of prosperity, when it was the "in thing" to have two names, or when people began having business away from home. When most common Europeans adopted surnames during the Industrial Revolution, there was a natural tendency to take the name of the important family in their area. It is therefore not surprising that although these "spaghetti" haplotypes are too distant from those in the central clusters to be related in the recent past, they usually belong to the same haplogroup. The spaghetti can be considered a collection of nonpaternity events with respect to the core cluster of the family.

The use of an unrelated surname could also have been an attempt to avoid persecution. Unfortunately, there are too many examples of this. Many Jews changed adopted

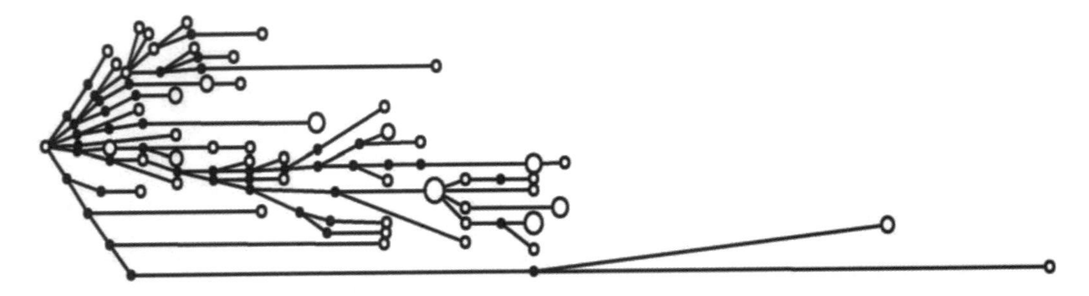

Figure 5. Cladogram of the Fitzpatrick DNA Study.

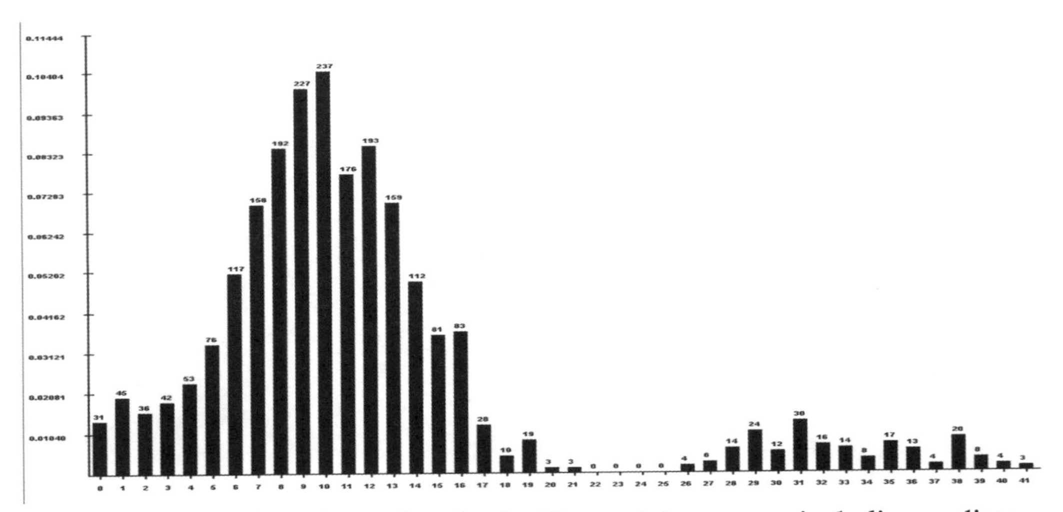

Figure 6. Pairwise mismatches for the Fitzpatrick surname including outliers.

Christian names during the Spanish Inquisition in the late 15th and early 16th centuries. More recently, during the Potato Famine of the mid-1800s, many Irish immigrants to the U.S. were forced to take the most menial and dangerous jobs that slaves were not allowed to do. Even highly educated Irish were sometimes forced to change their names to find decent work. The name of the architect who built Gallier Hall, the old city hall in New Orleans, for example, was not *Gallier*. It was *Gallagher.*[13]

The sixty-eight member Fitzpatrick study results show this type of structure (Figure 5). The cluster of haplotypes just below the middle of the figure represents Fitzpatricks from County Laois, Ireland, who descend from Bryan MacGiollaphadraig, Lord and First Baron of Upper Ossory (or his cook, as discussed above). There is another cluster of closely related haplotypes that originated in County Down, Ireland. Our theory is that the Fitzpatricks who belong to this cluster were originally Kirkpatricks who came from Scotland and who changed their name to Fitzpatrick to assimilate into the Irish culture.[14]

The graph of pairwise mismatches for the Fitzpatrick study is shown in Figure 6. A few pairs match closely, but the majority form a peak removed from the origin, yet not so far away that they can be classified as a collection of outliers. The peak represents mismatches between people in the spaghetti with each other and with members of the main group-pairs who are not closely related, but who probably belong to the same haplogroup. The small peak to the far left is due to outliers. These study members may not belong to the same haplogroup as the majority.

HOW MANY PEOPLE DO I NEED?

The more people in a study, the more you learn about your surname's history. If there is only one person in a study, he will have to look outside his study to assess relationships.

But for a study to produce meaningful results, it is not sufficient for it to include a large number of people. Its members should represent as diverse a cross section of people with the surname as possible. The more diverse the cross section, the more meaningful the results.

If a group includes many people from one branch of a family, but neglects others, it is not well-sampled. This could be unintentional. For very common surnames, obtaining participants from every branch of the family is impractical. Studies on common names are often restricted to people from a particular geographic location or those who are believed to descend from a specific ancestor.

Studies on rare surnames or studies that are restricted to a certain subgroup may remain small. Others will start off small but grow.

Figure 7a. Five randomly chosen Allred family members.

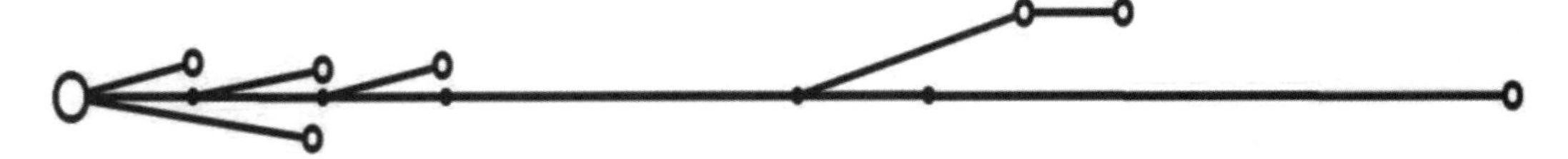

Figure 7b. Ten randomly chosen Allred family members.

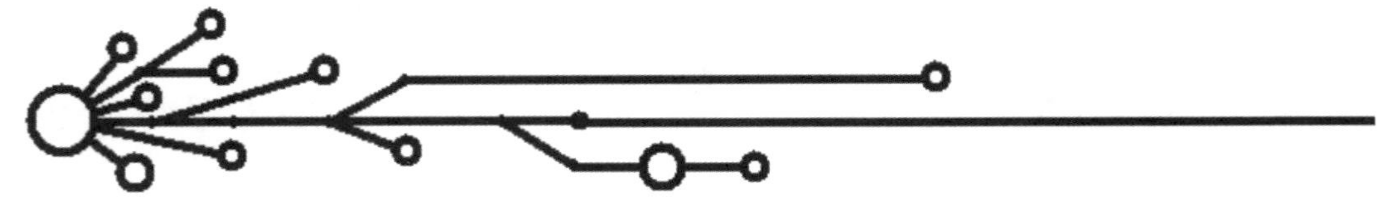

Figure 7c. All twenty-one Allred family members.

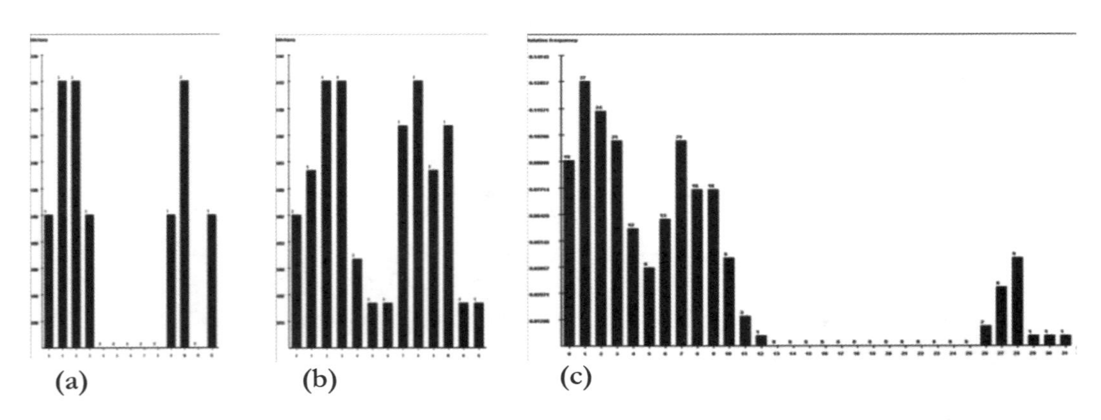

Figure 8. Plot of pairwise mismatches for Allred results based on (a) five, (b) ten, and (c) twenty-one members.

Is there a minimum study size necessary to draw meaningful conclusions? The answer is that it may be possible to identify some general features of a surname group starting with as few as five people, keeping in mind that any conclusions you draw are subject to change as more people are tested. At about twenty people, you can have more confidence that you are seeing the true structure of the name.

Figures 7a, b, and c show cladograms of the Allred surname study for five, ten, and twenty-one randomly chosen members. Figure 8 shows the corresponding plots of their pairwise mismatches. (The Allred study has only 21 members that belong to the study as of Fall 2005.)

Most of the Allreds appear to be closely related. With ten members, a star begins to take shape, becoming much better defined when all twenty-one members are included.

Figures 9a, b, and c show cladograms of the Fitzpatrick surname study for five, ten, and twenty-five randomly chosen members. Figure 10 shows the corresponding plots of their pairwise mismatches.

The Fitzpatrick surname shows high genetic diversity with only five members. The shape of the cladogram is that of a bush with many short to medium length branches. As new study members are included, the bush gets "bushier," with a few very long branches appearing that represent outliers. The haplotypes that are closest to that of the MRCA are separated from it by several mismatches. The cladogram and the plot of pairwise mismatches for all 68 members of the Fitzpatrick study are shown in Figures 5 and 6.

Figures 11a, b, and c show cladograms of the Baker surname study for five, ten, and twenty-five randomly chosen members. Figure 12 shows the corresponding plots of their pairwise mismatches.

The Baker study is the most genetically diverse, with branches radiating from the node of the MRCA. The study contains clusters of haplotypes near the MRCA, but also clusters that are genetically remote. The cladogram and plot of pairwise mismatches for the full study including all 62 Bakers are shown in Figures 3 and 4. For this common

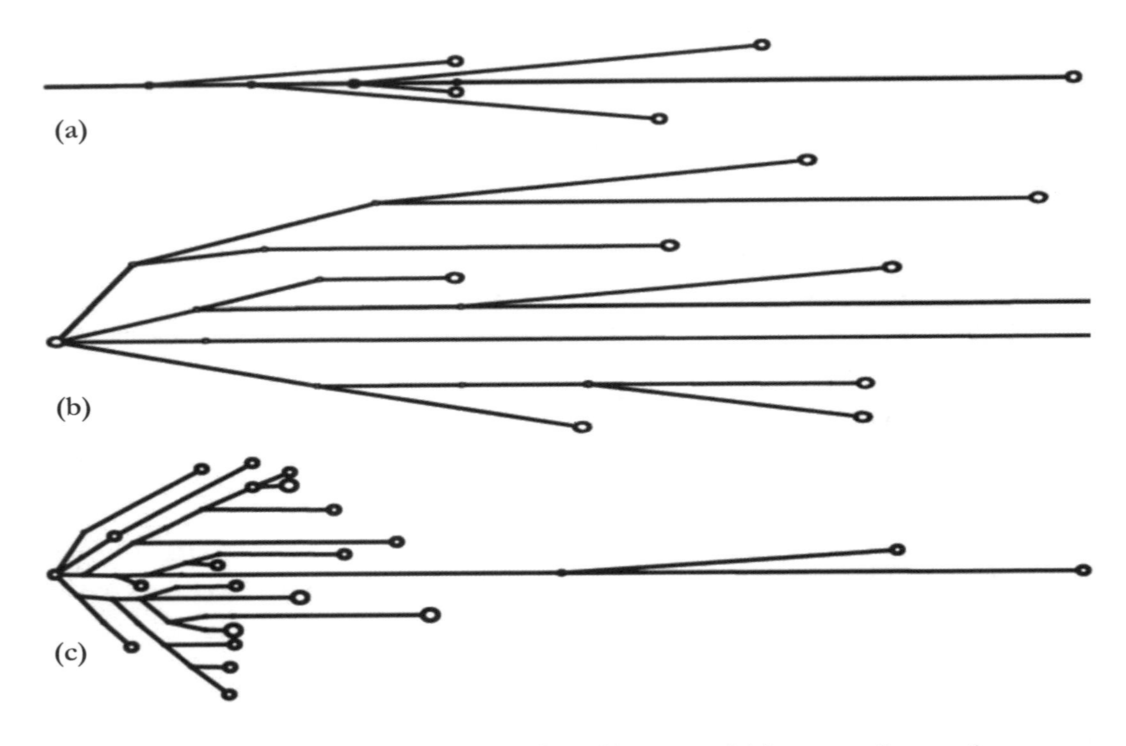

Figure 9. Fitzpatrick results of (a) five, (b) ten, and (c) twenty-five study members.

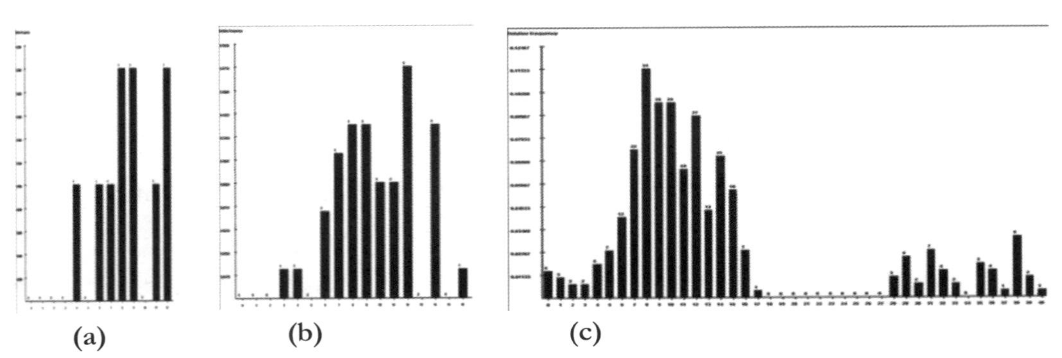

Figure 10. Fitzpatrick pairwise mismatches for (a) five, (b) ten, and (c) twenty-five members.

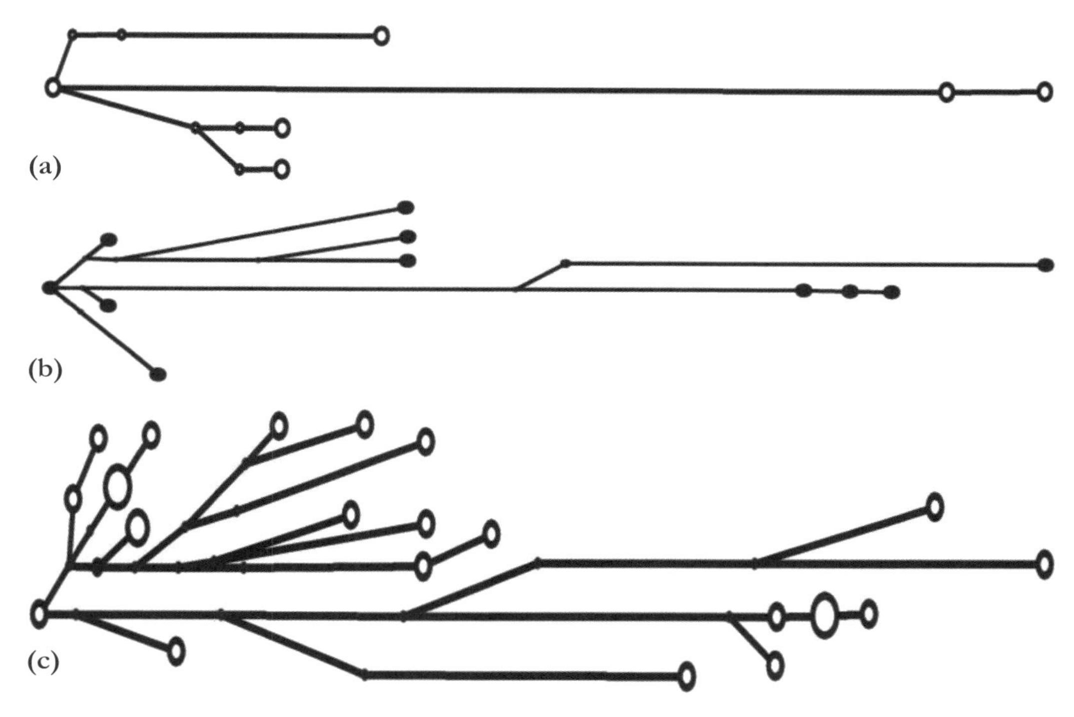

Figure 11. Baker results based on (a) five, (b) ten, and (c) twenty-five members.

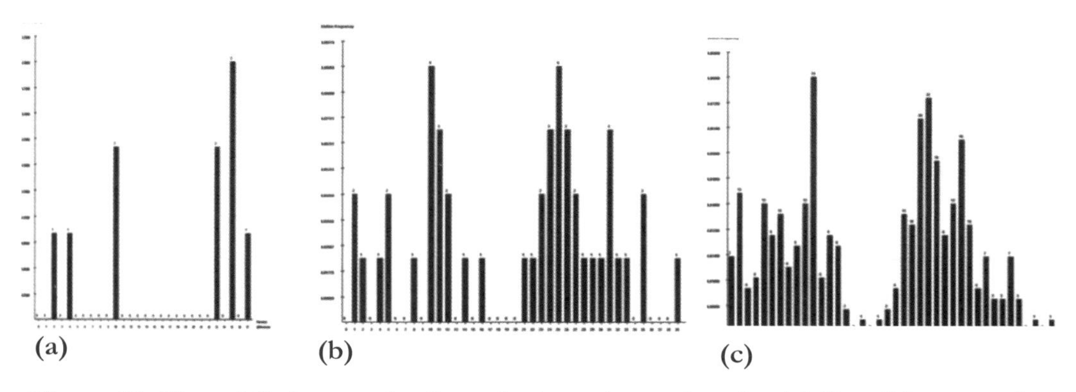

Figure 12. Plot of Baker results for pairwise mismatches for (a) five, (b) ten, and (c) twenty-five members.

Table 1. Allred study pairwise mismatch results.

No. of Members	Average Mismatches	Maximum Mismatches
5	4.2	12
10	5.0	12
19*	3.4	10

*21 group members minus 2 outliers

Table 2. Fitzpatrick study pairwise mismatch results.

No. of Members	Average Mismatches	Maximum Mismatches
5	8.6	12
10	9.2	16
25	9.2	17
65*	10.4	32

*68 group members minus 3 outliers

Table 3. Baker study pairwise mismatch results.*

No. of Members	Average Mismatches	Maximum Mismatches
5	15.3	27
10	18.8	39
25	17	39
62	16	38

*The Baker study is genetically diverse and could be considered a group of outliers.

surname, the peak representing high numbers of pairwise mismatches is just as large as the peak of closely related members. The sizes of the outlying peaks indicate the genetic diversity of a group. A study on a common name might include several different haplogroups.

Surnames can be analyzed by looking at the average number and the maximum number of pairwise mismatches. Stability of these variables as more study members are added can be compared. Tables 1, 2, and 3 show these characteristics for the three groups used here as examples.

Analysis of pairwise mismatches might not be straightforward because of the influence of outliers can bias calculations. At the very early stages when a study has only a few members, identifying outliers so that they can be eliminated from the analysis may be difficult, even for rare surnames. As a study grows, outliers may become obvious for rare or intermediate names. For a study of a common name containing several different haplogroups, however, the elimination of outliers might be difficult at any stage because of the genetic diversity of the surname.

As its membership grows, a study's results at various stages can be compared for the same conditions, for example, with or without outliers or using the same number of markers. When a study includes ten members, both the average and maximum number of pairwise mismatches begin to stabilize, and usually do not change very much as more members are added.

REFERENCES–WHAT CAN DNA TELL ME ABOUT MY SURNAME?

1. http://www.sca.org/heraldry/laurel/names/namehist.html.
2. http://genealogy.about.com/library/authors/ucboey2a.htm.
3. http://reference.allrefer.com/encyclopedia/N/name.html.
4. http://scripts.ireland.com/ancestor/magazine/surname/.
5. http://www.obcgs.com/LASTNAMES.htm#nicknames.
6. http://www.s-gabriel.org/index.html.
7. As printed in *Select Documents of the English Lands of the Abbey of Bec*, edited for the Royal Historical Society by Marjorie Chibnall, Ph.D., Camden Third Series, Volume LXXIII, London, Offices of the Royal Historical Society, 1951.
8. http://www.wordorigins.org/histeng.htm#early.

[9] http://www.shakespeare-online.com/biography/.

[10] webpages.charter.net/motuahina/kuauhau.html.

[11] http://www.ocnus.net/cgi-bin/exec/view.cgi?archive=59&num=15436.

[12] C. Fitzpatrick, *Forensic Genealogy*, Rice Book Press, p. 202 (2005).

[13] www.hgghh.org/user/WhyUsePrimarySources.pdf.

[14] Seamus Brennan, private communication.

SNPS, CLADES, AND HAPLOGROUPS

SNPS AND THE HAPMAP

The SNPs found on the Y-chromosome that are used for genetic genealogy are not all there is to the SNP story. SNPs occur on every chromosome. There are an estimated 10 million of them in nuclear DNA, one SNP for about every 100 to 300 nucleotides. SNPs account for 90% of the genetic variation in humans. The Y-SNPs used for genetic genealogy are not the only ones on the Y-chromosome either. They just happen to be the ones that are used by population geneticists and genetic genealogists. And don't forget that SNPs occur in mtDNA too.

SNPs (pronounced *snips)*, are locations in DNA where an alteration of a single base pair has occurred in at least 1% of a population. A SNP is also known as a polymorphism. It is possible that the same alteration appears in less than 1% of another population. In this case, it is considered to be a mutation. In other words, "One man's mutation is another man's polymorphism." (See Sickle Cell - Good News/Bad News on p. 35).

Most SNPs are benign, that is, they occur in noncoding or junk-DNA. A few SNPs are known to be located within genes, and to alter the protein the gene produces. One of the genes associated with Alzheimer's disease, called apolipoprotein E (ApoE), can contain two SNPs that may occur in three possible combinations known as E2, E3, and E4. Each combination differs by one DNA base, causing the protein product produced by a gene to differ by one amino acid. Research has shown that if a person inherits a copy of E4, he is much more likely to develop the disease than someone who inherits a copy of E2.[1]

The International HAPMAP Project is a multi-nation effort to catalog the SNPs that exist in nuclear DNA. While the Human Genome Project established the sequence of nucleotide pairs contained in chromosomal DNA, the HapMap Project is concerned with identifying human DNA sequence variations. The HAPMAP Project was started in 2002.[2] It has participants from the U.S., Japan, the U.K., Canada, and Nigeria, each of whom is responsible for characterizing a different region of the genome. SNP data is stored in the dbSNP database maintained by the U.S. National Center for BioTechnology Information (NCBI).[3]

The HAPMAP being produced by this project is useful in identifying genes associated with genetic disorders. While two unrelated individuals have 99.9% of their genome in common, the HAPMAP Project is concerned with identifying the remaining 0.1% which accounts for all their genetic differences, including their risk for developing disease and their variation in drug response.[2] Correlating a SNP sequence with the incidence of a disease in a population helps to determine the region in the genome where the genes associated with the disease might occur.

The HAPMAP can be reduced in complexity by taking advantage of the fact that SNPs that are close to each other on the genome are often inherited together. One SNP might show a variation within a population while other nearby SNPs do not. A set of related SNPs is called the haplotype for that region of the genome. The SNPs that uniquely identify each haplotype are called the "tags" for that region. Most of the common haplotypes for each region of the genome occur in all populations but in different frequencies. By creating a map of tag-SNPs, the HAPMAP can be reduced from 10 million SNPs to about 200,000 to 1 million tag SNPs, much easier to navigate than a map of the 3 billion base pairs that comprise the human genome.

For example suppose that four people have the genetic sequences:[2]

Chromosome #1 SNPs

	Region 1	Region 2	Region 3
Person 1	AACA**C**GCCA	TTCG**G**GGTC	AGTC**G**ACCG
Person 2	AACA**C**GCCA	TTCG**A**GGTC	AGTC**A**ACCG
Person 3	AACA**T**GCCA	TTCG**G**GGTC	AGTC**A**ACCG
Person 4	AACA**C**GCCA	TTCG**G**GGTC	AGTC**G**ACCG

If all the common bits are removed from each person's genome, a map of *only* the SNPs might look like:

Haplotype #1	...GTGAAAGTA**CGG**TTCAGGTA...
Haplotype #2	...TTGATTGCG**CAA**CAGTAACA...
Haplotype #3	...GCCGATCTG**TGA**TACTGGCG...
Haplotype #4	...TCGATTCCG**CGG**TTCAGACA...

(The same nucleotides are underlined in each case. The SNPs on either side of them are present on other parts of chromosome #1 not shown above.) Haplotype #1 has unique SNPs values at three locations:

Haplotype #1	GTGAA	A	GT	A	**CGG**TTCAGG	T	A
Haplotype #2	TTGAT	T	GC	G	**CAA**CAGTAA	C	A
Haplotype #3	GCCGA	T	CT	G	**TGA**TACTGG	C	G
Haplotype #4	TCGAT	T	CC	G	**CGG**TTCAGA	C	A

These are the tag-SNPs used to identify Haplotype #1. The other three haplotypes would be distinguished by different tag-SNPs.

Note that according to this definition, both male and female DNA include many different haplotypes and that a man or woman belongs to a different haplogroup for each region of the genome being studied.

HAPLOGROUPS AND CLADES

Back to genetic genealogy…

For the rest of this section we'll stick to the genetic genealogy definition of haplogroup associated with Y- and mtDNA-SNPs that are used for genetic genealogy. SNPs provide information that is useful in deriving the origins and migration patterns of indigenous peoples. However, SNP analysis is only an indicator of the development of present populations.[4] There are many possible explanations on how a group could have been formed and could have arrived at a geographical location. It is important to correlate theories of population development based on SNPs with archeological and anthropological evidence.

SNPs by definition are markers with only two values, meaning that they have mutated only once in human history. Each SNP can be traced to a single common ancestor where the SNP first appeared. The earliest SNPs can be traced to a Y-chromosome Adam who lived between 60,000 and 90,000 year ago[5] and a Mitochondrial Eve who lived about 150,000 years ago.[6] They are the most recent common ancestors of all humans along the direct male and females lines. Y-Adam and Mitochondrial Eve lived thousands of years apart, and were not the only male and female living at the time, but they are the only male and female whose direct male and female lineages survive to the present day. For example, other male lineages might have died out because other men had no children or had only daughters. Other female lineages might have died out because other women had no children or only sons.

The focus of geographic SNP research is indigenous populations, defined as (1) a cultural group (and its descendants) who have an historical continuity or association with a given region, or parts of a region, and who formerly or currently inhabit the region and who (2) existed before the colonization of the region, or who exist independently or largely isolated from the influence of later immigrants.[7] Due to the mobility of modern populations, many people do not live in the place where they were born, or have not lived in their present location for many generations. Studying nonindigenous populations would not give useful information in tracing very deep ancestral roots.

Y-SNP HAPLOGROUPS

Male populations have been organized into a few hundred major haplogroups and subgroups based on their Y-chromosome SNP mutations.[8] They are named according to the hierarchical nomenclature system of Phylocode[9] developed by the International Society for Phylogenetic Nomenclature and by the Y-Chromosome Consortium.[10]

The 19 major haplogroups are designated by uppercase letters of the alphabet including the letter Y which is used to signify the superhaplogroup that includes haplogroups **A** through **R**. The names of nested haplogroups alternate alpha and numeric characters.[11] For example, haplogroup **R1b** is the most common haplogroup present in European populations. It is believed to have expanded into and recolonized Europe 10,000 to 12,000 years ago after the last glacial maximum.

Because the greatest genetic diversity is in Africa, and because SNPs that occur in Africa are the most widely shared by populations in other parts of the world, most researchers agree with the single origin theory (also called the Out of Africa theory) This theory says that all modern humans evolved in Africa and then emigrated in several waves over the last 100,000 years, ultimately replacing earlier homo species.[12]

The geographical distributions and frequencies of SNPs allow us to construct the hierarchy of SNP haplogroups shown in the phylogenetic tree in Figure 1.[13] Haplogroups **A** and **B** exist only in Africa today. They are the oldest haplogroups. Haplogroup **B** exists in all parts of Africa, but is most common among the pygmy tribe.[14] The **CR** superhaplogroup (including Haplogroups **C** through **R** and their common ancestor) is distinguished by the **M168** mutation, which was carried during three migrations into the other parts of the world by some of the people with this mutation, while

Figure 1. Phylogenetic tree of Y-chromosome SNP haplogroups.[13] The CR super haplogroup is differentiated by mutations M168 and P9.

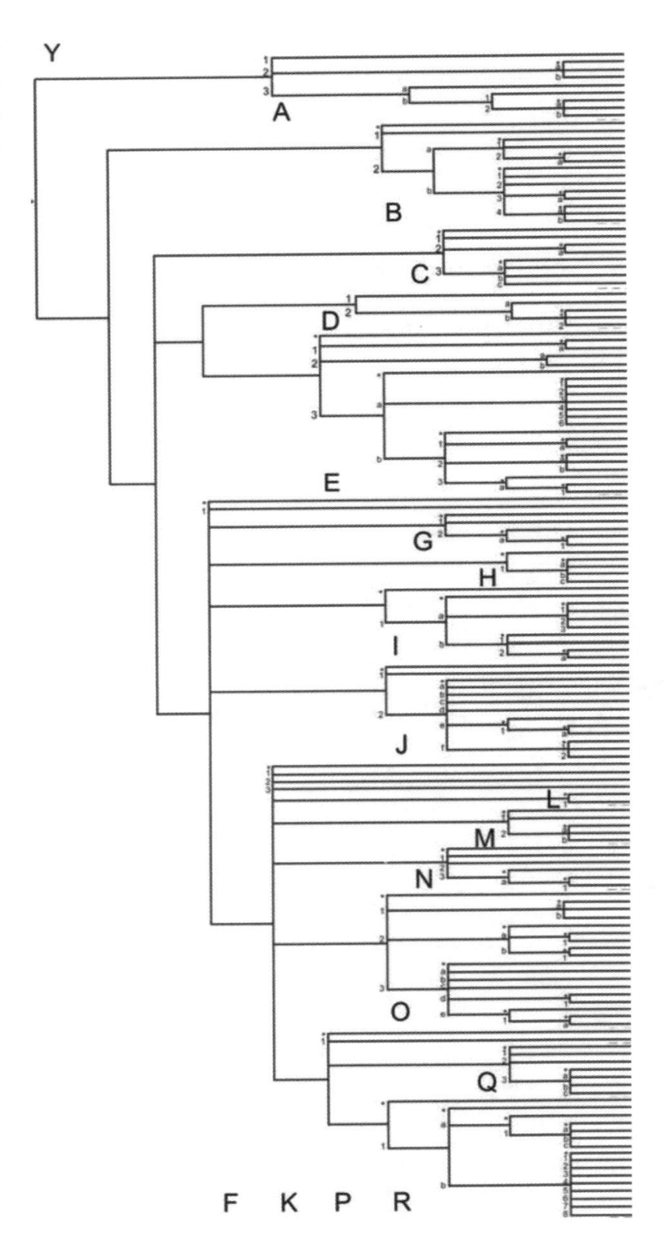

others remained behind. The three migrations happened at different times and are characterized by three different groups of mutations: (**YAP**, **M145**, **M203**) in Haplogroups **D** and **E** found in Africa and southeast Asia, (**RPS4Y**, **M216**) in Haplogroup **C** found in Asia, the South Pacific, and in low frequencies in Native American populations, and (**M89**, **M213, P14**) in Haplogroups **F** through **Q** in all areas of the world.[15]

By following the Y-phylogenetic tree from a larger branch down through one of its smaller branches, it has been possible to reconstruct the history of the smaller branch. For example, following the path from Superhaplogroup **CR** through nested groups **F**, **K**, **P**, **R**, **R1**, and **R1b**, a chronology has been derived for subgroup **R1b**. It was apparently formed from a group that emigrated from Africa 50,000 years ago, eventually settling in Europe in the last 30,000 years. See Table 1.[16]

Results of SNP analysis are usually reported as either a positive or a negative match to a certain haplogroup, based on the SNP observed. Because SNPs are expensive to test, yet yield valuable historical and geographical information, there has been much research into predicting SNP haplogroups using STR markers. The probability that an individual belongs to a haplogroup can be estimated from his haplotype without SNP testing.

The mutation rate for Y-SNPs used in population studies is very low. The average mutation rate for the Y-STR markers used in genetic genealogy is much higher. Over a long period of time, while some members of a haplogroup experienced additional SNP which defined new subgroups, their STR markers changed much more rapidly, so that each haplogroup today contains many haplotypes. Some haplogroups have drifted apart in their STR composition, while others still overlap.

The method used for haplogroup estimation based on STR results depends on the characteristic haplotypes found in each haplogroup. There is an online haplogroup predictor at https://home.comcast.net/~whitathey/predictorinstr.htm.[17] Given a haplotype, the predictor returns the probability of belonging to various haplogroups or subgroups. (See Figure 2).

Table 1. R1b SNP History[16]

Y-SNP	Haplogroup or Subgroup	YBP*	Migration Route
M94	BR	---	In Africa
M168	CR	50k	=> Middle East
M89	F	45k	=> Southwest Asia
M9	K	40k	=> North Central Asia
M45	P	35k	=> North West Asia
M207	R	--	In North West Asia
M173	1	30k	=> Europe
M343	b	--	In Europe

*Years before the present.

One of the exciting frontiers in genetic genealogy is the identification of new SNPs that tease apart the known haplogroups into ever smaller and more recent populations. At present, STRs indicate relationships over the past few centuries, while SNPs indicate connections over the last few millennia. Even the smallest SNP-defined groups include large segments of the population with a large distribution of STR haplotypes. This has led to ambiguity in estimating someone's haplogroup from his STR results. Subdividing the smallest branches of the Y-phylogenetic tree with new SNPs will identify more recent populations, and will reduce the number of STR haplotypes represented by these populations.[18]

MTDNA AND CLADES

Mitochondrial DNA is inherited exclusively along the female line. Males and females inherit mtDNA, but only females pass it to their children. If mitochondria, as it is widely believed, were once independent bacteria that colonized the ancestors of modern day membrane-bound cells, it is not surprising that mitochondria come equipped with their own DNA.

Enter the STR Values for the Haplotype to be Tested:

393	390	19	391	385a	385b	426	388	439	389i	392	389ii
13	25	14	11	11	13	12	12	12	13	14	29

458	459a	459b	455	454	447	437	448	449	464a	464b	464c	464d
18	9	11	11	11	25	15	18	30	15	16	16	17

460	GATAH4	YCA2a	YCA2b	456	607	576	570	CDYa	CDYb	442	438
11	12	19	23	17	0	0	0	0	0	12	12

How Well Does the Haplotype Fit the Following Haplogroups (Scale of 0 to 100):

E3a	E3b	G	I1a	I1b	I1c	J2	N	Q	R1a	R1b
3	5	1	0	8	1	9	1	43	16	59

No. of Markers
32

Values above 50 indicate a good fit with the haplogroup. Values between 20 and 50 indicate a fair fit.
Values below 20 indicate an unlikely membership in that haplogroup.

Ver 1.20

Figure 2. Haplogroup predictor. See https://home.comcast.net/~whitathey/hapest/hapest.htm[15]

Unlike nuclear DNA that is divided into chromosomes containing a total of 3 billion base pairs, the mitochondrial genome is a circular structure only 16,569 base pairs long. While nuclear DNA contains an estimated 30,000 genes, mtDNA contains only 19 genes.

Mitochondrial DNA is divided into the coding and the control regions. A mutation occurs somewhere in the coding region on the average of every 5,138 years. A mutation occurs in the control region on the average of every 20,180 years. The lower number of SNPs in the coding region is useful for defining deep female ancestry. The higher number of SNPs in the control region allows more detailed analysis of maternal lineages. Mutations can take the form of deletions, substitutions, or additions of a base pair.

Continued on p. 157

ANNA ANDERSON

In 1920 an unidentified woman was dragged out of Berlin's Landwehr Canal after an apparent suicide attempt. At first Fräulein Unbekannt (Jane Doe) refused to talk or give her name, but within a short time she was tentatively recognized as Tsar Nicholas II's youngest daughter Anastasia. Her claim as the sole survivor of the brutal execution of the Tsar and his family in 1918 by the Bolshevik revolutionary government was the source of one of the most compelling controversies of the 20th century. By the early 1990s, evidence examined by an international community including forensic anthropologists, forensic ondontologists, handwriting experts, historians, and experts in photograph analysis resulted in a close to 100% certainty that Fräulein Unbekannt,[1] commonly known as Anna Anderson, was indeed Anastasia. Later, the results of DNA analysis showed an almost certainty that she was an insane Polish factory worker Franziska Schanzkowska. In the end, DNA evidence prevailed.

Anna Anderson lived a life of public controversy, the most seriously considered of all those claiming to be survivors of the Romanov execution. The huge fortune allegedly left behind in foreign banks by the Tsar, coupled with the contradictory statements of the Soviet government on the whereabouts of the family or their bodies, set the stage for more than a few imposters to come forward over the decades claiming to be a missing Romanov. None was taken more seriously than Anna Anderson.

Anna had many who supported the legitimacy of her claim, including members of the imperial Russian family. Anna had many detractors who gave sworn affidavits that she was an imposter, including members of the imperial Russian family. On December 22, 1968, she married Dr. John Manahan in Charlottesville, VA. It was a marriage of convenience. She was 72, he was 49. Towards the end of her life, she became increasingly senile, dying on February 12, 1984. She was cremated the same day.

Continued on p. 157

Anna Anderson (cont)

It was only later in September of the same year that the concept of DNA fingerprinting was realized. Over the next few years, the power of DNA analysis was demonstrated for an ever increasing number of applications.

In 1991 bodies exhumed from an unmarked grave in Siberia were identified through DNA testing as those of the murdered Tsar and some, but not all, of his family. The bodies of only three of their five children were in the grave. By elimination, forensic examination showed that the two missing children were the Tsar's son Alexis and his youngest daughter Anastasia.

The mystery of Anna Anderson took on a new significance.

Continued on p. 158

Continued from p. 155

The control region, also called the D-loop, is the region where mtDNA starts the unzipping process during replication. The D-loop accounts for 7% of the mtDNA genome. Two sequences in the control region are usually tested for genetic genealogy: Hypervariable Region 1 (**HVR1**, including base pairs 16024 through 16383) and Hypervariable Region 2 (**HVR2**, including base pairs 57 through 372). Dr. Bryan Sykes used the **HVR1** region in his book *The Seven Daughters of Eve* to show that 98% of all European women descend from seven clan mothers who lived within the last 50,000 years. Mitochondrial DNA clades are labeled similarly to Y-chromosome haplogroups. Major mtDNA clades are designated by uppercase letters of the alphabet, with the names of nested subclades alternating numbers and lower case letters.

Dr. Sykes' book is a popular representation of a body of scientific research that has identified a total of 36 major clades according to their characteristic mtDNA-SNPs. Of these, the seven clades identified by Dr. Sykes as **H** (Helena), **J** (Jasmine), **U** (Ursula), **T** (Tara), **K** (Katrina), **X** (Xenia), and **V** (Velda) (in order from largest to smallest) originated in different parts of Europe at different times. Each of them accounts for a different fraction of the current European population, as shown in Table 2.

Anna Anderson (cont)

Anna Anderson had been cremated, but fortunately a tissue sample was discovered from a colon biopsy she had had at the Martha Jefferson Memorial Hospital four years before her death. After a 10-month legal battle that involved a documentary maker, a Russian immigrant association, the granddaughter of Dr. Botkin the Tsar's physician, the Martha Jefferson Hospital, Baron Ulrich von Gienanth who claimed to be the sole survivor of four executors Anna had named in her will, several law firms, and at least one other woman claiming she was Anastasia, Dr. Peter Gill, director of the Molecular Research Center of the Home Office Forensic Science Service arrived in Charlottesville and sliced specimens of Anna's colon that had been preserved in five paraffin blocks at the hospital.[2] The samples were brought to the British Ministry of Defense atomic-research facility at Aldermaston, the laboratory that had done the DNA analysis on the remains of the Tsar and his family.

Continued on p. 159

An additional clade **I** (Iris) with origins in the Ukraine accounts for most of the remaining 2% of Europeans not covered by the others.[19] Other current populations are grouped into other major clades–American Indians have four (**A**, **B**, **C**, and **D**), for example. A diagram of the population flows of the major world clades is found in Figure 3.[20]

The genetic diversity of mtDNA sequences found in Africa supports the evidence provided by Y-SNPs that Africa was where humans originated. As with Y-haplogroups, the most widely shared mtDNA-SNPs indicate the deepest genetic ancestry. Like Y-SNPS, mtDNA-SNPs are powerful tools for studying important aspects of population growth and geographical expansion. Through comparison of the mtDNA-SNPs appearing in different populations, a chronology for the expansion of humans into different parts of the world has been constructed and coordinated with such influences as climatic changes and shifts from hunter-gathering to farming. Mitochondrial DNA chronology can be adjusted to agree with archeological evidence for population expansions, such as a greater number of settlements, higher rates of refuse deposition,

Anna Anderson (cont)

In October 1994, the results were announced: DNA from the tissue taken from Anna Anderson did not match that of the Russian Imperial family nor did it match DNA from a blood sample obtained from Prince Philip, a distant relative of Empress Alexandra. However, the results were similar to those obtained from a German farmer named Carl Maucher, a collateral descendent of Franziska Schankowska,[3] a Polish peasant who had disappeared from Berlin at about the same time Fräulein Unbekannt had been saved from drowning in the canal.

The results were met with triumph by Anna's camp of detractors, and with dismay from her camp of supporters. There were claims that the hospital had switched biopsy samples, that perhaps other samples had been substituted while en route to England,[4] that something had caused a mix-up in the samples. None of these claims was found to have any substance. The Martha Jefferson Hospital had additional samples of Anna's tissue securely stored elsewhere in the hospital. It would have

Continued on p. 160

Table 2. Major European mtDNA Clades.

Haplogroup	% Europeans	YBP*	Origin	Notes
H	47	20k	South France	Largest
J	17	10k	Middle East	Second Largest
U	11	45k	Greece	Oldest
T	9	17k	Tuscany	--
K	6	15k	N. Italy	--
X	6	25k	Rep. Georgia, Asia	--
J	5	17k	N. Spain	--
I	2	26k	Ukraine	Smallest, Second Oldest

*Years before the present.

Anna Anderson (cont)

been impossible for someone without authorized access to switch them. Dr. Gill's results agreed with those obtained from different samples that had been sent to the Air Force Research Institute of Pathology in Bethesda. They also agreed with tests done on strands from a lock of hair found in an envelope labeled with Anna's name, stuck in an old book that had once belonged to John Manahan.

Through the decades, there had been hundreds of photographic comparisons between Anna Anderson and Anastasia. There had been graphologists testifying that Anna's handwriting was identical to Anastasia's. There was the coincidence that Anna had the same foot deformity that Anastasia had, and Anna's curiously detailed knowledge of the private lives of the Imperial family. There had even been experts who swore that Anderson's earlobes were the same as Anastasia's. But DNA evidence trumped them all.

1. Anna-Anastasia: Notes on "Franziska Schanzkowska", Peter Kurth, http://www.peterkurth.com/ANNA-ANASTASIA%20NOTES%20ON%20FRANZISKA%
2. *The Romanovs*, Robert K. Massie, Ballentine Books, 1996, pp. 194-226.
3. http://en.wikipedia.org/wiki/Anna_Anderson
4. http://www.freewarehof.org/manahans.html

and more colonization of areas that were previously unoccupied. The absence of mtDNA evidence of population growth can be used to indicate population distress caused by such influences as unfavorable weather conditions, disease, or pressure from other populations.[21] Such distress might create a population bottleneck where the number of individuals is greatly reduced, erasing the genetic diversity produced by earlier growth and expansion.

Mitochondria DNA was first sequenced in 1981 to create the Cambridge Reference Sequence (CRS).[22] It is the mtDNA profile of an arbitrary member of clade **H** (the clan Helena), the most common female clade found among people with Western Eurasian ancestry.[23] In a slightly modified form the revised CRS[24] is used as a master template against which to compare all other mtDNA profiles. The CRS for the **HVR1** and **HVR2** regions is shown in Table 3.

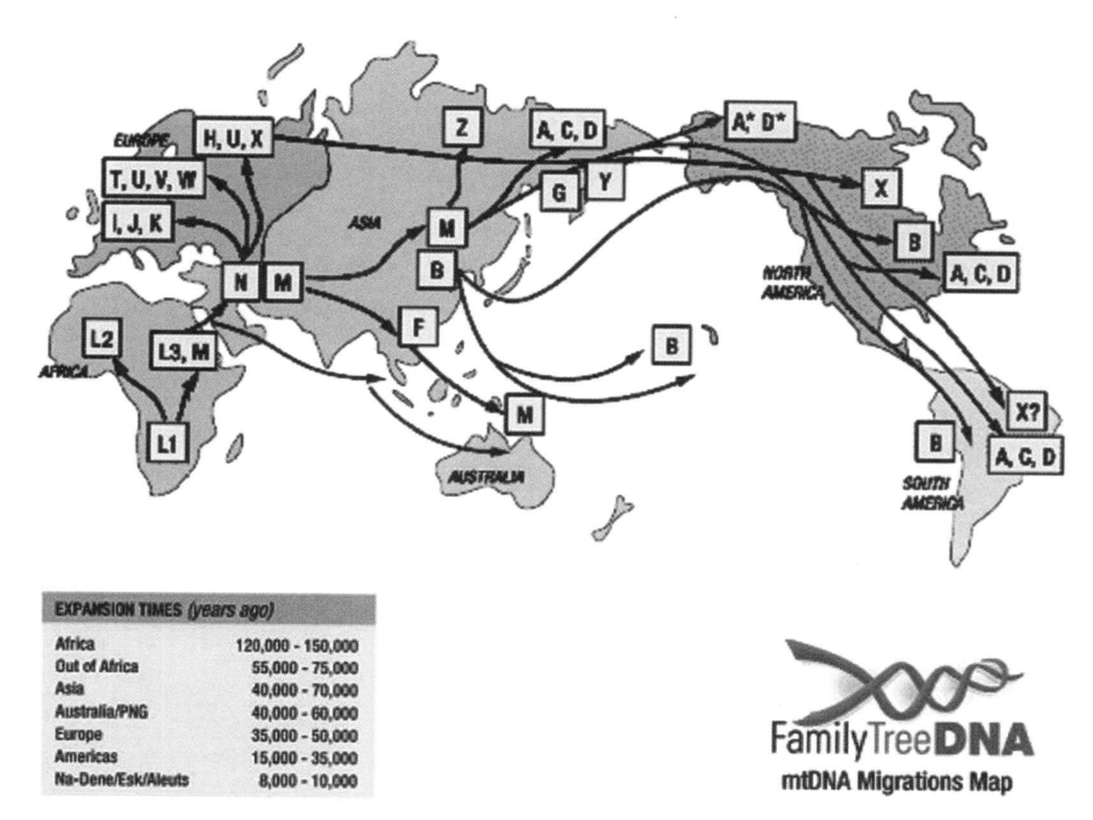

Figure 3. Major European mtDNA clades.

Results of individual mtDNA analyses are usually given in terms of deviations from the CRS. The sample results for the **HVR1** region shown in Table 4[25] indicate seven substitutions relative to the CRS. The number indicates the position in the mtDNA genome where the substitution occurs, with the letter indicating which base has been substituted into the mtDNA. For example, the notation **16129A** stands for the substitution of adenine (**A**) for thymine (**T**) in position **16129** in the CRS. The CRS sequence from position 16126 through position 16130 reads TATTG. Including the first mutation in Table 4 changes the sequence to TATAG.

Table 3. The Cambridge Reference Sequence

HVR1 Reference Sequence							
16010▼	16020▼	16030▼	16040▼	16050▼	16060▼	16070▼	16080▼
ATTCTAATTT	AAACTATTCT	CTGTTCTTTC	ATGGGGAAGC	AGATTTGGGT	ACCACCCAAG	TATTGACTCA	CCCATCAACA
16090▼	16100▼	16110▼	16120▼	16130▼	16140▼	16150▼	16160▼
ACCGCTATGT	ATTTCGTACA	TTACTGCCAG	CCACCATGAA	TATTGTACGG	TACCATAAAT	ACTTGACCAC	CTGTAGTACA
16170▼	16180▼	16190▼	16200▼	16210▼	16220▼	16230▼	16240▼
TAAAAACCCA	ATCCACATCA	AAACCCCCTC	CCCATGCTTA	CAAGCAAGTA	CAGCAATCAA	CCCTCAACTA	TCACACATCA
16250▼	16260▼	16270▼	16280▼	16290▼	16300▼	16310▼	16320▼
ACTGCAACTC	CAAAGCCACC	CCTCACCCAC	TAGGATACCA	ACAAACCTAC	CCACCCTTAA	CAGTACATAG	TACATAAAGC
16330▼	16340▼	16350▼	16360▼	16370▼	16380▼	16390▼	16400▼
CATTTACCGT	ACATAGCACA	TTACAGTCAA	ATCCCTTCTC	GTCCCCATGG	ATGACCCCCC	TCAGATAGGG	GTCCCTTGAC
16410▼	16420▼	16430▼	16440▼	16450▼	16460▼	16470▼	16480▼
CACCATCCTC	CGTGAAATCA	ATATCCCGCA	CAAGAGTGCT	ACTCTCCTCG	CTCCGGGCCC	ATAACACTTG	GGGGTAGCTA
16490▼	16500▼	16510▼	16520▼	16530▼	16540▼		
AAGTGAACTG	TATCCGACAT	CTGGTTCCTA	CTTCAGGGTC	ATAAAGCCTA	AATAGCCCAC		
HVR2 Reference Sequence							
70▼	80▼	90▼	100▼	110▼	120▼	130▼	40▼
CGTCTGGGGG	GTATGCACGC	GATAGCATTG	CGAGACGCTG	GAGCCGGAGC	ACCCTATGTC	GCAGTATCTG	TCTTTGATTC
150▼	160▼	170▼	180▼	190▼	200▼	210▼	220▼
CTGCCTCATC	CTATTATTTA	TCGCACCTAC	GTTCAATATT	ACAGGCGAAC	ATACTTACTA	AAGTGTGTTA	ATTAATTAAT
230▼	240▼	250▼	260▼	270▼	280▼	290▼	300▼
GCTTGTAGGA	CATAATAATA	ACAATTGAAT	GTCTGCACAG	CCACTTTCCA	CACAGACATC	ATAACAAAAA	ATTTCCACCA
310▼	320▼	330▼	340▼	350▼	360▼	370▼	380▼
AACCCCCCCT	CCCCCGCTTC	TGGCCACAGC	ACTTAAACAC	ATCTCTGCCA	AACCCCAAAA	ACAAAGAACC	CTAACACCAG
390▼	400▼	410▼	420▼	430▼	440▼	450▼	460▼
CCTAACCAGA	TTTCAAATTT	TATCTTTTGG	CGGTATGCAC	TTTTAACAGT	CACCCCCCAA	CTAACACATT	ATTTTCCCCT
470▼	480▼	490▼	500▼	510▼	520▼	530▼	540▼
CCCACTCCCA	TACTACTAAT	CTCATCAATA	CAACCCCCGC	CCATCCTACC	CAGCACACAC	ACACCGCTGC	TAACCCCATA
550▼	560▼	570▼					
CCCCGAACCA	ACCAAACCCC	AAAGACACCC					

The sample results for the **HVR2** region shown in Table 5 indicate substitutions, additions, and deletions of bases relative to the CRS. Additions are indicated by the number of the position in the genome after which the addition takes place, followed by a decimal point, a digit, and the initial of the added base. If the numeral after the decimal is a 1, the notation indicates the addition of a base directly after the position. If the digit is a 2, the base is the second addition after that position. For example, the notation **309.1C** means that cytosine (**C**) was added after position **309**, and **309.2A** means that an adenine (A) was added after the cytosine. A deletion is indicated by the number of the position followed by a minus sign, so that **523-** means that the base at position **523** has been deleted. The CRS from position **300** through position **310** reads AACCCCCCCT. Including the two insertions at position **309,** the sequence reads AACCCCCCC**CA**T.)

Table 4. Sample mtDNA HVR1 results.

16129A	16223T	16391A	16085G	16172C	16311C	16519C

Table 5. Sample mtDNA HVR2 results.

73G	**199C**
203A	**204C**
250C	**263G**
309.1C (Insertion of a C after location 309)	
309.2A (Insertion of an A after the previous insertion.)	
315.1C (Insertion of a C after location 315.)	
455.1T (Insertion of a T after location 455.)	
523- (Base pair at location 523 is missing.)	
524- (Base pair at location 524 is missing.)	

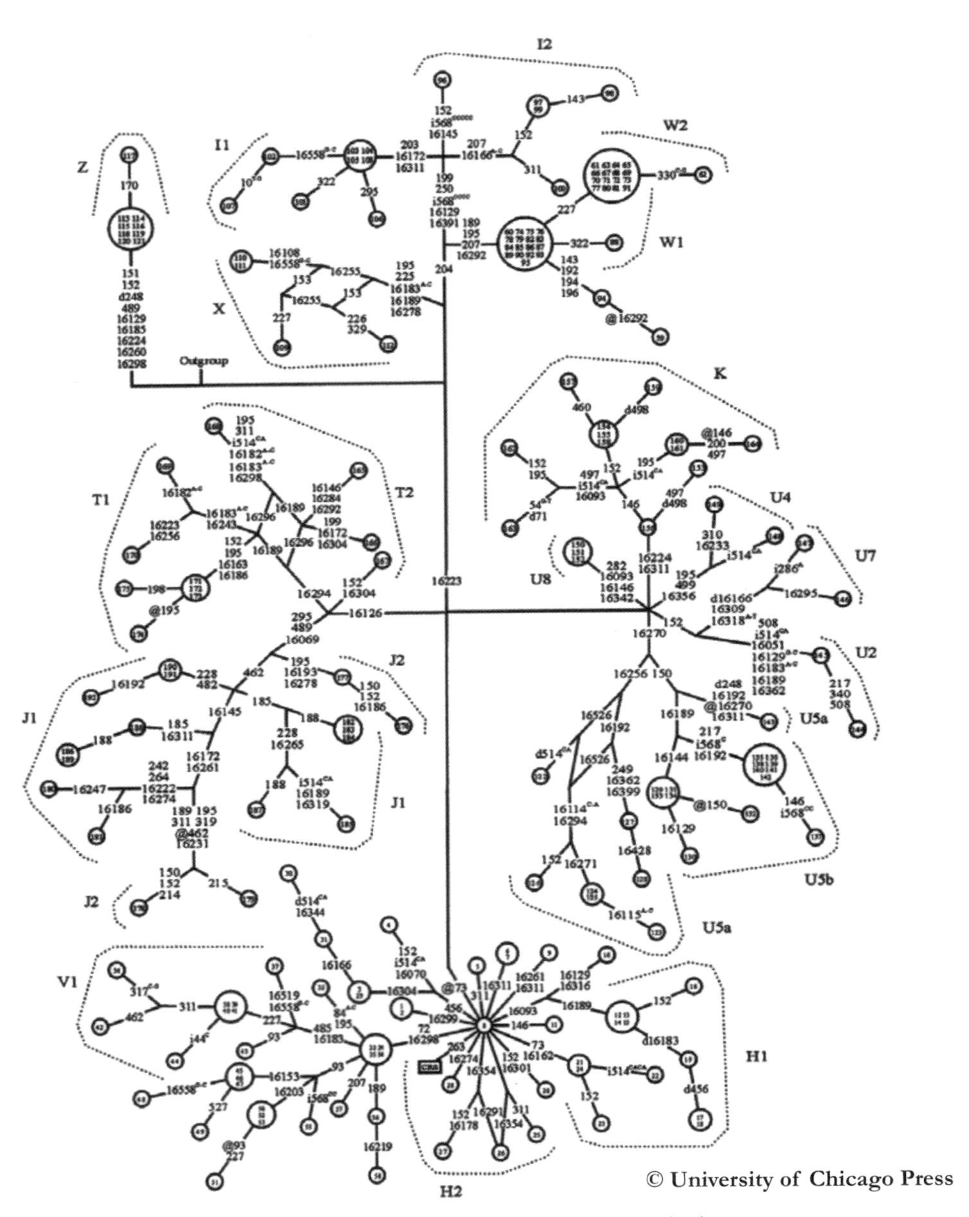

Figure 4. SNPs defining European clades.

The clade an individual belongs to is determined by comparing his mtDNA mutations to those characteristic of the various mtDNA haplogroups. In the examples given in Tables 4 and 5, the **HVR1** mutations at **16129**, **16233**, **16391**, and the **HVR2** mutations at **199**, **204**, and **250** are indications of Clade **I**. The additional mutations in **HVR1** (**16311C**, **16172C**) and in **HVR2** (**203A**) are indications of the mtDNA subclade **I1**. The remaining mutations could be random mutations that either define an as yet unidentified subsubclade of Clade **I** or are random mutations that are not shared by a large enough population to define a subsubclade. The relationship between these mutations and Clade **I** can be seen in the upper left hand area of Figure 4.[26]

REFERENCES–SNPS, CLADES, AND HAPLOGROUPS

1. Human Genome Project Information, SNP Fact Sheet, http://www.ornl.gov/sci/techresources/Human_Genome/faq/snps.shtml#snps1.
2. http://www.hapmap.org/abouthapmap.html
3. NCBI, Single Nucleotide Polymorphisms, http://www.ncbi.nlm.nih.gov/SNP/index.html
4. P. Underhill, "Human Y-chromosome phylogeography: clues to the population origin, affinity, migration, subdivision, and amalgamation," *Racialization and Populations, Society and Science*, October 23-24, 2002, Bethesda, MD.
5. http://en.wikipedia.org/wiki/Mitochondrial_eve
6. http://en.wikipedia.org/wiki/Y-chromosomal_Adam
7. "Indigenous Peoples", http://en.wikipedia.org/wiki/Indigenous_peoples
8. The Y-Chromosome Consortium, "A Nomenclature System for the Tree of Human Y-Chromosomal Binary Haplogroups", *Genome Research*, **12**(2), 339-348, February 2002.
9. http:www.ohiou.edu/phylocode/art9html and http://miketaylor.org.uk/dino/faq/s-class/phylectic/
10. http://ycc.biosci.arizona.edu/nomenclature_system/results.html
11. The Y-Chromosome Consortium, "A Nomenclature System for the Tree of Human Y-Chromosome Binary Haplogroups," **12**(2), 339-348, February 2002.
12. Hillary Mayell, "Documentary Redraws Human Family Tree", National Geographic News, January 21, 2003.
13. http://ycc.biosci.arizona.edu/nomenclature_system/fig1.html
14. http://www.roperld.com/YBiallelicHaplogroups.htm
15. Ref. 2, Op. cit.
16. *The Journey of Man, A Genetic Odyssey* by Spencer Wells, p. 182. as quoted by Y-Chromosome Biallelic Haplogroups, http://www.roperld.com/YBiallelicHaplogroups.htm
17. https://home.comcast.net/~whitathey/predictorinstr.htm

18. Bennett Greenspan, Family Tree DNA, private communication.
19. Bryan Sykes, *The Seven Daughters of Eve*, Also see http://www.oxfordancestors.com/your-maternal.html and http://www.roperld.com/mtDNAdaughters.htm#haplogroups.
20. Used with the permission of Bennett Greenspan, Family Tree DNA.
21. J. F. O'Connell, "Genetics, archeology, and Holocene hunter-gatherers," *Proc. Natl Acad Sci USA*, **96**(19):10562-10563, Sept. 14, 1999.
22. Anderson, S., A. T., Bankier, B. R. Barrell, et al, "Sequence and organization of the human mitochondrial genome," *Nature* **290**:457-465, 1081.
23. http://www.oxfordancestors.com/glossary.html
24. Andrews, R. M., I. Kubacka, P. F. Chinnery, R. N. Lightowlers, D. M. Turnbull, N. Howell, "Reanalysis and revision of the Cambridge reference sequence for human mitochondrial DNA," *Nat. Genet.*, **23**:147, 1999.
25. Information contained in Tables 4 and 5 and information included in the accompanying discussion are used with the permission of L. David Roper, http://www.roperld.com/mtDNA.htm
26. Figure 2 from Finnila, S., Lehtonen, M. S., Magamaa, K., "Phylogenetic Network for European DNA," *Amer. J. Hum. Gen.*, **68**(6):1475-1484 (2001).

THE LAST WOR(L)D

The Typewriter

In the Middle Ages, it was commonly believed that the earth was the center of the universe. It was not a coincidence that this corresponded with the Catholic Church's notion about the central importance of man to just about everything. Copernicus was the first to challenge this belief when he published his treatise *De revolutionibus orbium coelestium* (Nuremberg, 1543), arguing that the sun was near the center of the universe, near the center of motion for the earth and the other known planets. When his work was finally printed twenty years after it was originally formulated, it was greeted with respect, although more as an alternative mathematical way of describing celestial motion rather than truth.

In late 1609, Galileo pointed his newly invented telescope at the heavens and discovered four moons orbiting the planet Jupiter. He also pointed it at the sun, and saw sunspots. He was the first to observe that the planet Venus exhibits phases, much like the moon. These simple discoveries had profound implications not only for astronomy but for religion as well. If Galileo was right, and man could no longer claim to occupy the center of an imperfect physical universe. What did this imply about man's position in the eyes of God? There was a collision between literal interpretation of the Holy Scripture and scientific discovery. The Church was not pleased. Galileo was accused of heresy and placed under house arrest for the rest of his life. It was not until 1992, three hundred and fifty years later, that Pope John Paul II admitted that there were errors made by the theological advisors involved in Galileo's case. But he did so without admitting any mistake in accusing Galileo of heresy for his belief that the earth, like the other planets, revolves around an imperfect sun.

Galileo's 17th century discoveries set the stage for the identification over the next three hundred years of many other celestial "centers"–other moons orbiting other planets, spiral galaxies orbiting black holes, and even other planetary systems existing in these other galaxies. Man lost for good his physical importance as the center of the universe.

Yet even if there was nothing uniquely central about our physical position in the universe, couldn't we still make a claim that we, as human beings, had a unique intellectual importance in the scheme of things? We were after all the only intelligent life on this planet.

We could until the 1920s when Wolfgang Kohler, the psychologist, observed that chimpanzees not only used tools to reach food, but that they could also *make* tools to reach food. In the 1960s, Jane Goodall observed chimps doing so in the wild. Until then, it was thought that only humans were able to make tools. Did this mean chimpanzees were human? But couldn't we recover our position of importance by restricting the definition of human to depend on the use of language? What about emotions? And if that doesn't work, what about reasoning? Of course, we now know that dolphins have a language, elephants mourn their dead, and even cats and dogs reason in their own cat and dog ways. There are numerous stories of dogs saving their masters' lives in one way or another. We now know that even bees communicate the way to the nearest flower through a kind of bee-dance.

So we have again been dethroned in our self-specialness. We have had to admit we share the planet with lots of other forms of intelligent life.

But wait. Can't we claim that even though the Earth goes around the sun like all the other planets and even though there are countless stars that orbit centers called black holes that we can't ever hope to see, and even though we share the planet with other forms of intelligent life, can't we pride ourselves that we are the only life in the universe?

Guess what is going to happen next? Soon, I predict that we will discover that the universe is teaming with life–many forms of life. I believe that long ago, something happened in all the limitlessness of space and time that pulled together the most abundant elements–hydrogen, carbon, nitrogen, and oxygen, and *encoded* them into what we call life, and that furthermore this life somehow replicated itself, and that these replicas formed more replicas down many generations, until now, 13 ½ billion years later, the universe is *infected* with life.

I use the word encoded because the DNA we regard as the blueprint of life as we know it is in a way a code–the same chemicals adeline, guanine, cytosine, and thymine (AGCT), strung together and repeated in almost endless combinations, much as dots and dashes are repeated and recombined to form a message in Morse Code. It is as if the event that created the first life composed it on a typewriter with four keys instead of two- or three-dozen.

But because it is common in typing to strike the occasional wrong key and produce a typo, a copy is usually not quite the same as the original. As the life-typewriter continued to make copies of what it had created, there was the occasional miscopy that created a slightly new variation on the original theme. Most of these were literally dead ends, doomed never to reproduce themselves. But over such a long time, many miscopies successfully replicated, creating new forms of life that not only reproduced themselves but also mutated into other newer variations, populating the universe with a vast variety of living creatures.

So I have come to the conclusion that even if we are not the physical center of the universe, and even if we ride on one of nine planets orbiting only one of trillions of suns in billions of galaxies, and even if we are not the only intelligent life on this planet, and probably not even the most intelligent life in the universe, perhaps we can still be proud that we are a representation of that *one* original form of life, descendants of that *fundamental* manuscript that has been handed down and is now manifested in our very existence.

Or maybe not.

Colleen Fitzpatrick
November 2005

APPENDIX A
MRCA CALCULATIONS

To get a general idea of what is involved in an MRCA calculation, consider two members of a study group, A and B, who are one mismatch apart. If we could research the DNA profile of A's line generation by generation, starting at the present and moving backwards along his family tree, we would find that over the centuries, A's family's DNA mutated many times. (We could just as well use B's family line.) The most recent mutation accounts for the single difference observed between A and B. Since one has this mutation but the other does not, the mutation must have happened in some generation **G1** since their MRCA. For example, the mutation could have occurred in the grandchild of their MRCA.

If we continue to search further back, we will find a generation **G2** where A's family DNA shows a second mutation. Since both A and B have this mutation, it must have occurred in an ancestor they share, and their family lines must have branched off after **G2**. Their MRCA must have been born after **G2** but before **G1**.

Therefore, the probability that A and B's MRCA lived since any generation **G** is the probability that no more than one mutation happened since that generation. To calculate the probability they have a MRCA, we want to exclude the probability that two or more mutations happened over the intervening period of time.

To do this we resort to a truism: the probability of one mutation or zero mutation (no more than one mutation) added to the probability of more than one mutation is **1** (100% certainty). So to get the probability of more than one mutation happening, (that is, the probability that two people have a MRCA if they have one mismatch) we subtract the probability of no mutations (**P0**) and the probability of one mutation (**P1**) from **1**.

Probability of the MRCA within G generations based on 1 mismatch

= 1 - P0 - P1.

It's easy to see from this that an MRCA calculation involves one minus several terms. The number of terms is determined by the number of mismatches observed. Each successive term is associated with one more mismatch. In calculating the probability of an MRCA based on two mismatches, for example, you must consider the probability that intervening ancestors experienced zero, one and two mutations, but not three or more. In general, the probability of finding the MRCA since a generation **G** based on **k** mismatches is expressed by

Probability of the MRCA within G generations based on k mismatches = 1 - P0 - P1 - P2 … -Pk

where **Pk** is the probability that **k** mutations will happen since that generation.

TRANSMISSION EVENTS

In general, the more markers you are tested on, the more mismatches you should expect to find with someone else. It's not that the mutation rate increases when you add markers. It's that there are more markers that could have mutated since you shared a common ancestor with that person.

The probability of finding a mismatch between two people depends on how many transmission events have occurred since their MRCA, with a transmission event defined as passing one marker from a parent to a child. In Y-STR testing, this means a father to a son. The more markers you look at the more mutations you are likely to see. The more generations you consider, the more mutations you are likely to see.

The number of transmission events separating two people includes generations along both family lines since their MRCA. The number is calculated by counting the number of generations from the MRCA down to one of the people and then adding it to the number of generations from the MRCA down to the other person. Two brothers, for example, are separated by two transmission events per marker since their MRCA is their father, separated from each of them by one generation.

The total number of transmission events is the total number of generations along both family lines multiplied by the number of markers tested. (For simplicity it is usually assumed that there are the same number of generations along both lines, so that the total number of generations is twice the number along either line.)

Converting the number of generations to a time span involves assuming an average generation in years. This varies somewhat by culture and period of history. It has been conventional to assume an average of 25 years per generation in genealogical calculations. Recent research, however, indicates it's about 30 years for males in western cultures for the last few hundred years.[1]

THE BARE FACTS

The variables that are used in a MRCA calculation are:

n = the number of markers,
k = the number of mismatches,
G = the number of generations to the MRCA, and
r = the average mutation rate.

Because DNA testing companies normally regard the mutation rates of individual markers as proprietary, in performing your own MRCA calculations you must use the average mutation rate for the markers in the test panel.

An MRCA calculation involves inputting values for each of these variables into certain mathematical expressions, which provide you with the probability you share an MRCA with someone who lived within the last G generations, or:

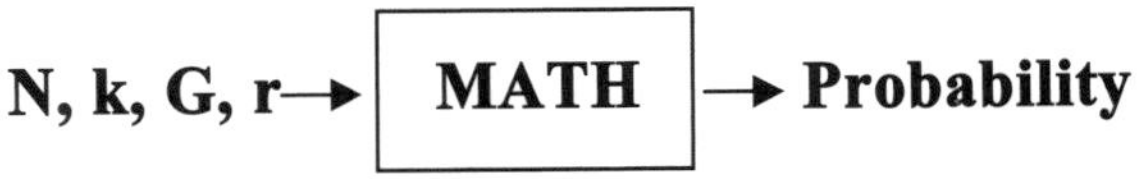

N, k, G, r → [MATH] → Probability

There are many sources for the box labeled "MATH" in the diagram. One of them it the Family Tree FTDNATiP™ calculator you can access from your MyFTDNA page if you are testing with Family Tree DNA. The FTDNATiP™ estimator automatically compares your results with someone else you match and calculates the probability you have an MRCA with that person within 100, 200, 300, 400, 500 and 600 years, based on your genetic distance from him. Although you cannot access them to use them yourself, the FTDNATiP™ calculator uses the individual mutation rates of the markers in the test panel to perform the calculations.

If you'd like to experiment with the results by varying the mutation rate or the number of markers, you can do so using Walker Moses' online MRCA calculator at http://www.moseswalker.com/mrca/calculator.asp?q=2. This calculator requires input for **n** and **r**, and will return either the number of generations to the MRCA as a function of probability or the probability versus the number of generations to the MRCA. If you want to observe more of the mathematical effects on the calculations of varying **n**, **G**, and **r**, the website www.forensicgenealogy.info offers downloadable spreadsheets that you can use to understand how the different variables influence the probability of a match.

The values for the four input variables come from different sources. The DNA test you take determines **n**; **k** is obtained by comparing your DNA results to someone else's; **G** is the number of generations in the past you are interested in investigating; and **r** is the average mutation rate, a constant provided by the DNA testing company for the test panel you have chosen.

Without going into the gory details of the math, there are a few common sense features of the calculations that are easy to understand:

1. If you perform your own MRCA calculation using an average mutation rate, but you observe a mismatch on one of the more rapidly changing markers in your test panel, your calculations will probably overestimate the time to your MRCA;

2. If you have a two-step mutation on a marker that changes very fast, it should be counted as a single mismatch. If you have a two-step mutation on a marker that changes very slowly, it should be counted as two mismatches.

3. If two people find they have a certain number of mismatches **k** when they test on **n** markers, and two other people who have tested on the same **n** markers find they have more than **k** mismatches, the first two people are probably more closely related than the second two people.

4. If you increase the number of markers you test on, but do not find any more mismatches with someone, your MRCA probably lived more recently than you thought. Testing on more markers allows you to have more mismatches and still be considered closely related to someone else.

5. If you use a higher mutation rate in interpreting your test results, the result will indicate that your MRCA probably lived in the more recent past than if you use a lower mutation rate for your calculations. Markers with a higher mutation rate are useful for investigating relationships in the more recent past; slower moving markers are useful for investigating relationships in the more distant past. A test panel will include both types of markers.

6. The more recently you are related to someone else, the less likely you will show mismatches with him.

MRCA CALCULATIONS AND BAYES THEOREM

MRCA calculations are based on Bayes' theorem that relates to the probability that an event happened in the past based on something that is observed in the present: in math terms, the probability that event **X** happened if event **Y** is observed, **P(X|Y)** or the probability of **X** given **Y**. This is a sensible approach to use for genealogy because DNA analysis tells us how many mismatches, **k**, occur on **n** markers between two individuals tested in the present (event **Y**) and, from this, we want to deduce the likelihood that they had a common ancestor who lived a certain number of generations, **G**, in the past (event **X**) or **P(G | k)**, the probability of **G** given **k**.

Another way of saying this is that the Bayesian approach to MRCA calculations depends on the probability of finding individuals within a population who have two characteristics in common. Not only do they show **k** mismatches on **n** markers, they also have a MRCA ancestor exactly **G** generations ago. There will be many people who have only one of these characteristics that will not qualify for the group. Many pairs of people will have **k** mismatches, but they will not have an MRCA exactly **G** generations ago, and many will have an MRCA exactly **G** generations ago, but will not exhibit **k** mismatches. If we find the fraction of people in the general population who have both characteristics, we can calculate the MRCA estimate.

In searching for this group, it does not matter which of the two attributes we investigate first. We could start with the fraction of pairs of people whose DNA profiles mismatch **k** times out of **n** markers, **P(k)**, and then select from them only those with an MRCA exactly **g** generations ago **P(G | k)**. The total probability of finding those with both characteristics is the product of these two probabilities, or **P(G | k)*P(k)**. (Call this the forward direction.) Or we could do it the other way around. We could find the fraction of the people in the general population who share an MRCA exactly **g** generations ago **P(G)** and then select from them, narrowing the group down to only those who show **k** mismatches on **n** markers **P(k | G)**. The total probability of finding those with both characteristics is the product of these two probabilities, or **P(k | G)*P(G)**. (Call this the backward direction.) Either way, we will arrive at the same group of people who both have a MRCA exactly **G** generations ago and exhibit **k** mismatches out of **n** markers.

In other words, the forward probability is equal to the backwards probability, or

$$P(G \mid k)*P(k) = P(k \mid G)*P(G)$$

This is Bayes' theorem.

The term that we are after in this equation is the first one on the left, **P(G | k)**, which can be obtained from an evaluation of the other three terms in the equation as follows.

P(k) = the probability that two randomly selected individuals will show **k** mismatches out of **n** markers. This is a constant α that can be obtained empirically from DNA testing on a large number of randomly selected people. Alternatively, this constant can be determined by a mathematical process called normalization that involves summing the probabilities on the right hand side of the equation for all values of **k** and adjusting the sum so that it equals unity.

P(G) = the probability that two randomly selected individuals have a common ancestor exactly **G** generations ago. Since the amount of effort required to experimentally determine this factor would be prohibitive for now, it must be modeled through coalescence population theory. This theory indicates that, for a sample drawn from a population greater than about 250 individuals within which marriage occurs, **P(G)** can be set to unity without a significant effect on the calculations. Genealogical studies usually meet this criterion unless they involve very limited populations such as Stone Age tribes on isolated islands in the Indian Ocean. The reader is referred to Hudson[2] for more information on coalescence theory.

So for populations of more than 250 individuals, the equation reduces to

$$P(G \mid k) * \alpha = P(k \mid G) * 1$$

or

$$P(G \mid k) = P(k \mid G)/\alpha.$$

The remaining term to evaluate is **P(k | G)**, the probability that two individuals mismatch on **k** out of **n** markers given that they have a MRCA **G** generations ago. **P(k | G)** can be calculated using either the infinite allele or the stepwise model of mutations. The infinite allele model assumes that the possible outcomes of a mutation are infinite, that is, they do not occur through discrete changes in length. This implies that there is a negligible chance that two different loci will mutate in the same way, so that each mutation is unique. Using this model the only way two individuals can match on a marker is for that marker not to have changed. This model does not take into account the possibility of multiple mutations, including those where a mutation might have canceled itself out in a two-step process, changing in one direction the first time and then in the reverse direction the second time. The model also does not distinguish a multiple-step from a single-step mutation. The infinite-allele model for **k** mutations on **n** markers is nothing more than the binomial distribution representing the number of ways **k** items can be chosen from a pool of **n** total items.

The binomial expansion reduces to the Poisson probability distribution at low mutation rates and the large number of markers common to single-name studies:

$$P(G \mid k) = \frac{(2nG)!}{k!(2nG-k)!} r^k (1-r)^{2nG-k} = \frac{e^{-2nGr}(2nGR)^k}{k!}$$

where **R** is the mutation rate, **2nG! = 2nG * (2nG - 1) * (2nG - 2) * … * 1** and **t = 2nG** the number of chances for a mutation to occur.

The stepwise model of mutations assumes that the length of a marker changes in discrete steps, adding or subtracting one unit of length during each mutation. The stepwise model considers that a **+1** mutation, where the allele increases by one, has the same probability of a **-1** mutation, where the allele decreases by one. The stepwise model allows for multiple discrete mutations at a single marker, including pairs of mutations that cancel out. The stepwise model also includes the possibility that two separate family lines have experienced the same mutations on the same markers, maintaining a perfect match on these markers. Since the stepwise model includes the possibility of mutations that have reversed themselves, it gives a higher estimate for the number of generations back to the MRCA than the infinite allele model as long as enough time has gone by such that there is a non-negligible chance that more than one mutation has occurred.

In practice, MRCA calculations depend on many empirical parameters that have not yet been accurately determined, such as mutation rates of many of the individual markers. Stepwise calculations are much more difficult than infinite-allele calculations. Although the stepwise approach is closer to the physical phenomena being modeled than the infinite allele method, both models give the same results within the margin of error in cases where the probability of more than one mutation is low, or where the number of generations **G** < 0.25/**R**. For the mutation rate of **R**= 0.004, **G** = 62.5 generations.[3] Considering a generation to be 25 years, this is equivalent to about 1563 years, well outside the range of conventional genealogy.

REFERENCES–APPENDIX A, MRCA CALCULATIONS

1. M. Tremblay and Helene Vesina, "New Estimates of Intergenerational Time Intervals for the Calculation of Age and Origins of Mutations," *Am. J. Hum. Genet.*, **66**:651-658, 2000.
2. Hudson, R. R., "Gene genealogies and the coalescence process," *Oxford Surveys in Evolutionary Biology*, D. J. Futuyama and J. Antonovics, eds., Oxford University Press, Oxford, pp. 1-44.
3. B. Walsh, "Estimating the Time to the Most Recent Common Ancestor for the Y chromosome or Mitochondrial DNA for a Pair of Individuals," *Genetics*, **158**:879-912 (June 2001).